D0889889

PICTURES, IMAGES, AND CONCEPTUAL CHANGE

SYNTHESE LIBRARY

STUDIES IN EPISTEMOLOGY,

LOGIC, METHODOLOGY, AND PHILOSOPHY OF SCIENCE

Managing Editor:

JAAKKO HINTIKKA, *Florida State University*

Editors:

DONALD DAVIDSON, *University of Chicago*

GABRIËL NUCHELMANS, *University of Leyden*

WESLEY C. SALMON, *University of Arizona*

VOLUME 151

JOSEPH C. PITT

Department of Philosophy, University of Pittsburgh

PICTURES, IMAGES, AND CONCEPTUAL CHANGE

An Analysis of Wilfrid Sellars' Philosophy of Science

D. REIDEL PUBLISHING COMPANY

DORDRECHT : HOLLAND / BOSTON : U.S.A.

LONDON : ENGLAND

Library of Congress Cataloging in Publication Data

Pitt, Joseph C.
  Pictures, images, and conceptual change.

  (Synthese library ; v. 151)
  Bibliography: p.
  Includes indexes.
  1. Science—Philosophy.   2.   Change.   3.   Sellars,
Wilfrid.   I.   Title.
Q175.P55   501   81–8705
ISBN 90–277–1276–X   AACR2
ISBN 90–277–1277–8 (pbk.)

Published by D. Reidel Publishing Company,
P.O. Box 17, 3300 AA Dordrecht, Holland.

Sold and distributed in the U.S.A. and Canada
by Kluwer Boston Inc.,
190 Old Derby Street, Hingham, MA 02043, U.S.A.

In all other countries, sold and distributed
by Kluwer Academic Publishers Group,
P.O. Box 322, 3300 AH Dordrecht, Holland.

D. Reidel Publishing Company is a member of the Kluwer Group.

TABLE OF CONTENTS

PREFACE

In this essay I am concerned with the problem of conceptual change. There are, needless to say, many ways to approach the issue. But, as I see it, the problem reduces to showing how present and future systems of thought are the rational extensions of prior ones. This goal may not be attainable. Kuhn, for example, suggests that change is mainly a function of socio-economic pressures (taken broadly). But there are some who believe that a case can be made for the rationality of change, especially in science. Wilfrid Sellars is one of those.

While Sellars has developed a full account of the issues involved in solving the problem of conceptual change, he is also a very difficult philosopher to discuss. The difficulty stems from the fact that he is a *philosopher* in the very best sense of the word. First, he performs the tasks of analyzing alternative views with both finesse and insight, dialectically laying bare the essentials of problems and the inadequacies of previous proposals. Secondly, he is a systematic philosopher. That is, he is concerned to elaborate a system of philosophical thought in the grand tradition stretching from Plato to White-head. Now with all of this to his credit, it would appear that there is no difficulty at all, one should simply treat him like all the others, if he indeed follows in the footsteps of past builders of philosophic systems. Yes, but that is precisely the problem, for how do you treat the great philosophers of the past? You either examine the systems for consistency, evaluate the merits of the presuppositions utilized and conclusions derived, or attempt to fit the views into the developing history of philosophy.

Trying to determine the consistency with which Sellars works out his ideas presents a set of problems of its own. To begin with, he argues to his own conclusions dialectically, subtly playing off real and imagined points as they meet his needs. The matter is complicated by the often frustrating promissory notes. It is not that he fails to eventually make good on them, it is just difficult to put all the pieces together at any one time, since they emerge in apparent random order.

Given those difficulties, we can try a different approach and attempt an evaluation of his presuppositions and conclusions. But given the piecemeal fashion by which the system is played out, the exact force of the presuppositions is hard to determine. Furthermore, the system is incomplete and premature judgment tends to lead to unhappy results when you pick up the next issue of some journal.

Finally there is the option of seeing Sellars in historical garments. This is well nigh impossible for a man who is deeply indebted to Kant, Wittgenstein (early *and* late), Carnap, Pritchard, Aristotle and Peirce. If we try to label him in terms of schools of philosophical thought — even limited to the twentieth century — we are somewhat stymied, for he is a systematic analytic Wittgensteinian who is also a pragmatic scientific realist.

While I am generally sympathetic to the types of considerations to which Sellars attends, there are two areas in which I see problems. First, while like Sellars, I do not believe that there is conflict between what he calls the scientific and the manifest images of man, I nevertheless find his account of their relation inadequate. Second, his theory of probability, on which his account of the development of new theories rests, suffers from several conceptual difficulties. In dealing with these difficulties there is obvious need for knowing how these issues bear on his general system. Such global considerations, however, are beyond the scope of this work. My objective is limited. I am concerned with Sellars' vision of the future of scientific development and the manner in which he works out these ideas using the notion of a conceptual framework. The view I defend in the long run owes much to Sellars' views, but incorporates a number of ideas contrary to his concerning the relation between these frameworks and the world. My skeptical conclusions are more the result of my own inability to put content on the bones of conceptual framework language than a clear rejection of Sellars' position. Also, I hope to show that what he wants can be done in a somewhat more straightforward fashion.

Chapter One provides an overview of Sellars' philosophy of science, touching on what I take to be key areas of interest for our purposes. Specific attention is paid to Sellars' Peircean view that science is a process which will culminate in a complete theory at some future point.

One of Sellars' more important contributions to the discussion of change lies in his insistence on the rationality of scientific inquiry. In Chapter Two the sense in which inductive inference is presupposed as a rational mode of reasoning is discussed. Sellars' views here are compared with Goodman's and Quine's.

Utilizing the presuppositions argued for in the discussion of induction, Chapter Three focuses on Sellars' theory of probability. It is to provide the mechanism whereby new information is generalized and worked into the theoretical structure of science formalized in terms of a theory. How theories function is the topic of Chapter Four.

In the final chapter I argue against Sellars, not merely because of the problems in the theory of probability, but because of complications in the conceptual apparatus he generates to handle the sense in which science *evolves*. Traditional attempts to explicate the rationality of science are based on the idea that some steady progress is being made toward a full account of the physical universe. Changes in theories are judged rational or not in terms of the contribution such changes make toward achieving this end. While Sellars disagrees with more traditional approaches on many points, he shares this assumption and I find it suspect. But to deny that change is rational is not to deny it is reasonable. My argument concludes that change in science can be described as reasonable and, hence, justified if it can be explained. The broad extension of this idea asserts the non-rationality of conceptual change in general, but likewise does not deny its intelligibility.

Various points covered in my discussions here have appeared elsewhere in somewhat altered form. Part of Chapter III appeared in *Philosophy Research Archives*, Vol. II as 'Wilfred Sellars' Theory of Probability'. A large section of Chapter IV appeared in *Dialectica*, Vol. 34 as 'Hempel versus Sellars on Explanation'. In addition I drew liberally from the general presentation to formulate my introduction to *The Philosophy of Wilfrid Sellars: Queries and Rejoinders*.

Needless to say, the friends, colleagues and teachers whom I must thank for their varied assistance and encouragement are many, too many to all be given proper acknowledgment. A special word of thanks is due Robert Butts for his efforts to help me understand what is important; Frank Neumann and Bob Cohen have been continuing sources of encouragement; and Betty Q. Davis is gratefully noted for her cheerful typing and editing. But this work is due mostly to Donna's patience and constant support, and it is to her that I give my final thanks.

Newport, Virginia   JOSEPH C. PITT
15 November 1980

ACKNOWLEDGEMENTS

'The Language of Theories' by Wilfrid Sellars from *Current Issues in The Philosophy of Science*, edited by Herbert Feigl and Grover Maxwell. Copyright © 1961 by Holt, Rinehart and Winston. Reprinted by permission of Holt, Rinehart and Winston.

'The Theoreticians Dilemma' by Wilfrid Sellars. Reprinted with permission of Macmillan Publishing Co., Inc. from *Aspects of Scientific Explanation* by Carl G. Hempel. Copyright © 1965 by The Free Press, a Division of Macmillan Publishing Co., Inc.

Fact, Fiction and Forecast by Nelson Goodman, published by Harvard University Press, 1955, reprinted by permission.

'Philosophy and the Scientific Image of Man' by Wilfrid Sellars, reprinted from *Frontiers of Science and Philosophy*, Robert G. Colodny, editor. Published in 1962 by the University of Pittsburgh Press. Used by permission.

Wilfrid Sellars, 'Theoretical Explanation', *Philosophy of Science: The Delaware Seminar, Vol. 2,* edited by Bernhard H. Baumrin (New York: John Wiley – Interscience), pp. 329–331.

The Structure of Science by Ernest Nagel, by permission of Ernest Nagel and Hackett Publishing Company, Inc.

'Scientific Realism or Irenic Instrumentalism' by Wilfrid Sellars in *Boston Studies in the Philosophy of Science,* Vol. 2, edited by R. Cohen and M. Wartofsky (Dordrecht: D. Reidel), by permission of Humanities Press Inc., New Jersey 07716, and by permission of D. Reidel Publishing Company.

Science and Metaphysics by Wilfrid Sellars (London: Routledge & Kegan Paul, 1968), reprinted by permission of Humanities Press Inc., New Jersey, 07716, and by permission of Routledge and Kegan Paul.

'Some Reflections on Language Games' by Wilfrid Sellars, reprinted from *Philosophy of Science*, Vol. 21, 1954, copyright © The Williams & Wilkins Co., Baltimore.

'Induction as Vindication' by Wilfrid Sellars, *Philosophy of Science*, Vol. 31, No. 3, 1964. Reprinted by permission of The Philosophy of Science Association.

PICTURES AND TELEOLOGY

1. SCIENCE, PHILOSOPHY, AND CHANGE

Change is one of the fundamental facts with which the scientist deals. The philosopher is also concerned with change. His worries, however, are not those of the scientist. The scientist's concern can be located in his efforts to construct theories to explain and predict with increasing adequacy the changes which occur in the physical world. The philosopher worries about the conceptual requirements for inquiry entailed by the efforts of the scientist.

In particular, there are two different, but related, sets of worries concerning the questions of change to which the philosopher directs his attention: the conceptual tools with which the scientist conducts his inquiry, i.e., theories, and the nature of the scientific enterprise itself. In the first case he examines the viability of specific modes of explicating certain concepts employed in explaining the events of the world, e.g., 'law' and 'evidence'. In the second case he evaluates, catalogues, and explicates scientific inquiry itself. Part of this job involves elaborating a conceptual framework within which science is to be viewed. Since scientific inquiry is a dynamic developing process, a minimal condition of adequacy for the philosophical framework is that it be able to handle the conceptual changes which are part of the development of science.

While scientists always use theories in their systematizations, they sometimes abandon one theory in favor of another. Given the results of more traditional positivistic analyses of the relation between theory and evidence,[1] there arises the problem of establishing rational grounds for rejecting one theory and accepting another.

Now this is not a question about the psychology of decision-making, nor does it concern only the means of justifying an acceptance or rejection. Rather, it is first and foremost a question about the conceptual prerequisites for such action. To rationally reject a theory requires that the decision take place in a conceptual framework which encompasses more than the theory itself. This framework must extend beyond the theory in order for an evaluation of the theory to be possible. Without the possibility of such an evaluation we eliminate any chance to be rational with regard to the decision. It must also be a large enough framework to permit alternative paths of inquiry when the inadequacy of a given course of action becomes evident. Otherwise, the shift to another theory as another way of organizing data about the world could not be described as rational. That is, no reasons could ever be given to explain why the choice of theory was a good one, or even why the old theory was a bad one.

The problem of explicating change in science, where 'change' refers here to science and not the world, precipitates a question concerning the proper characterization of science. Is science a process of a product? Not all share the working assumption here that science should be analyzed as a process of inquiry.[2] The usual approach to the process of science involves examining the activities of scientists discovering, developing, and testing hypotheses and theories. But doubts arise as to whether that is the proper object of investigation for the philosopher. The proper object, it is argued, is the finished tested theories and the criterion for testing these theories; in other words, science is best discussed in terms of its product.

Basically, this conflict concerns the proper method of analysis. Should we seek criteria for evaluating the products of science or should we seek criteria by which we can rationally seek alternatives to the present answers and ways of giving answers? Distinguishing between the logic of justification and the logic of discovery to characterize the objectives of these two quests, the latter question concerns the possibility of a logic of discovery. Is there a justifiable method for *generating* new theories, which method would in some way guarantee the usefulness of those theories? Is the process of scientific inquiry justifiable, or must we limit our efforts to justifying only the product?

But, on the other hand, is the product of science justifiable? Wilfrid Sellars agrees with the positivists that it is. However, on his account the dichotomy between process and product must be eliminated in favour of a teleological account of scientific inquiry. While Sellars has argued against the positivist approach to the justification of theories, nevertheless, he believes that there is a sense in which the product of science can be justified. Moreover, he sees in the manner of justifying the product of science a way for the process of inquiry to be justified in turn.

There has been, however, a close identification of the idea of the process of science and the logic of discovery with psychological issues of creativity. In what is oddly called *The Logic of Scientific Discovery*, Karl Popper gives short shrift to the problems of the scientific process. The topic of Popper's discussion is justification, the justification of whatever it is we may dream up dozing in front of a fireplace. In his mind this comes to the problem of testing our hypotheses, and so his real concern lies in the logic of testing, not with the far richer notion of discovery.

Popper believes that the "logical analysis of scientific knowledge" requires a distinction between questions of fact and questions of validity. ([31], p. 31) He classifies all questions surrounding the inventing of new ideas as questions of fact. While they may be of interest to psychology, Popper considers them irrelevant to the problem of the validity of knowledge and dismisses them. The question of validity for Popper comes to the issue of testing.

Accordingly I shall distinguish sharply between the process of conceiving a new idea, and the methods and results of examining it logically. As to the task of the logic of knowledge — in contradistinction to the psychology of knowledge — I shall proceed on the assumption that it consists solely in investigating the methods employed in those systematic tests to which every new idea must be subjected if it is to be seriously entertained. ([31], p. 31)

As a result of these distinctions, Popper's 'logic of discovery' turns into a logic of justification.[3] We "discover" what it is we are justified in claiming by carefully articulating the means of testing hypotheses. That is, the sense of 'discovery' here is rather limited. Claims for a logic of discovery *usually* call to mind rules for generating new theories, hypotheses and generalizations, not for testing old ones. Such systems could easily be confused with a proposed inductive logic, however, and Popper emphatically claims that "in my view there is no such thing as induction". ([29], p. 40)

But even with an inductive logic in hand we would still be hard pressed to develop a logic of discovery, where this entails a means for proposing new ideas. Logic is concerned with validity. The most we should expect from an inductive logic is a systematic analysis of the relations between the old concepts and the new ones which we create. We could quite legitimately have an inductive logic and still not have a means of creating new hypotheses. How people come up with hypotheses is a question for psychology, a question of fact.

Because our pre-analytic notions concerning a logic of discovery appear to require something stronger than merely an inductive logic, 'something else' with psychological overtones, and because questions of change in science appear to be associated with discovery, the analysis of change has suffered. Not only has the analysis of change suffered, but the analysis of the logic of justification has run into severe difficulties. With the possibility of a logic of discovery in disrepute, and given the close association of change with discovery, the early argument over the proper method to take in analyzing science has temporarily ended in favor of those who consider the proper object of philosophical analysis to be the structure of the product of scientific inquiry, theories, as well as methods of testing.

In general, the distinction between the logics of justification and discovery has resulted not only in the dismissal of one half of the distinction, but also in the ossification of philosophical inquiry with respect to science. The denial of the possibility of a logic of discovery *seems* to imply the denial of the possibility of analyzing the scientific process. The argument being that if there is no logic of discovery then there cannot be an analysis of conceptual change, i.e. the framework and prerequisites for changing theories.

However, science must be analyzed as a process, not merely as a product. This is not to deny that the analysis of theories and the problems of justification are important, nor to suggest they be left for another day. But it is only in the context of looking at more than individual theories that the problems of both the justification and the analysis of theories can be completely answered.

Methods of analyzing and explicating science differ with respect to how they view both the goal and structure of science. In this respect, we can distinguish between logic and method. Logic refers to patterns of inference. Hence, the *logic* of science can be described as deductive, provided we limit ourselves to describing the formal relations between sentences in a reconstruction of a finished and complete theory. The *method* of science, on the other hand, concerns strategies for gathering data, formulating and testing hypotheses and generalizations, proposing, testing, and accepting laws and theories.[4]

My arguments here, as an exercise in methodology, are one step removed from the use of 'method' above, where the study of scientific methodology would be the descriptive study of the activities of scientists. I am not concerned *directly* with what scientists do. This task belongs to history, psychology and sociology. Here I examine the claims of certain philosophers concerning the goal, structure and function of science in general and of scientific theory in particular. Moreover, the criterion of adequacy for the viability of one or another method of analyzing science is not whether science actually looks like the picture painted by either of these views. This would be inadequate since those issues center on how to characterize or reconstruct science. Instead, the criterion to be used here is epistemological. Does the proposed view of scientific inquiry yield a viable account of the product science produces: explanatory knowledge?

Sellars conceives of science as an evolving process culminating in a theory which will replace our prescientific conception of the nature of the universe. In this way he includes both sides of the discussion. The product is clearly identified and the manner is captured in which the process of inquiry both feeds on and is governed by its goal. His theory of change in science, programatically stated in 'Philosophy and the Scientific Image of Man', attempts to redeem the analysis of the development of science from the psychological quagmire to which it was relegated when placed in the domain of the logic of discovery.

2. IMAGES

For Sellars, the development of science is a specific type of refinement of what he calls "the framework in terms of which, to use an existentialist turn of phrase, man first encountered himself – which is, of course, when he came to be man". ([45], p. 6) In the original framework man described and explained the world around him anthropromorphically. The refinement of this image consisted in the gradual depersonalization of nature on the one hand, and, on the other, a systematizing of our empirical observations into generalizations. This latter enterprise Sellars calls the empirical refinement of the original image.

By empirical refinement, I mean the sort of refinement which operates within the broad framework of the image and which, by approaching the world in terms of something like the canons of statistical inference, adds to and substracts from the contents of the world as experienced in terms of this framework and from the correlations which are believed to obtain between them. ([45], p. 7)

This begins to sound very similar to the sorts of procedures which mark the *scientific* spirit. But while Sellars is willing to admit that the resulting refined image might be classified as the scientific image of man, since it "is not only disciplined and critical; it also makes use of those aspects of scientific method which might be lumped together under the heading 'correlational induction' ", ([45], p. 7) he prefers to call this the manifest image of man. He reserves the label of scientific image for the framework which results from the type of reasoning, peculiarly scientific, but not included under correlational methods, "namely that which involves the postulation of imperceptible entities, and principles pertaining to them, to explain the behaviour of perceptible things". ([45], p. 7)

On this view, growth in science involves a dialectical interchange between what he calls the correlational and postulational techniques. Sellars also believes, however, that "the notion of a purely correlational scientific view of things is both an historical and methodological fiction". ([45], p. 7) This constitutes an objection to a Humean analysis of science which insists on at least a partial defining of theoretical concepts in observational terms.[5] Such an analysis "involves abstracting correlational fruits from the conditions of their discovery and the theories in terms of which they are explained". ([45], p. 7)

The correlational fruits, i.e., empirical generalizations, are the products of the correlational methods. However, not only does Sellars believe that empirical generalizations are to be explained by theories, he also believes explanations do not result from deduction of generalizations from theoretical structures. Rather, we first explain the particular perceptibles using a theory. We explain the perceptibles by giving an analysis of what they really *are* in terms of the imperceptibles postulated by the theories. Then, because we now know what the perceptibles are, we can explain why their behavior is amenable to being described in terms of the empirical generalizations which seem to capture accurately their behavior. ([44], pp. 121, 123)

Thus, it is important to take note that when Sellars gives the criterion by which the scientific image is to be differentiated from the manifest, as above, the scientific image is characterized by the use of theories postulating unobservables to explain *observables*, not generalizations. However, it is equally important to realize that Sellars views the scientific image as an outgrowth of the manifest.

The point of this analysis might, however, be lost if we fail to consider the sense in which Sellars views the two images of man he has distinguished, manifest and scientific, as competing. For Sellars the job of the philosopher is to provide a steroscopic view of these two different views of the world. Because the scientific image explains the observable domain by postulating unobservable phenomena, we might expect a simple resolution of the issue in favor of the scientific, i.e., the real objects are those which science says exist. ([45], pp. 25–37)

But because man appears to reduce to a mere physical phenomenon when viewed only in terms of the scientific image, the issue becomes slightly more complicated. For man may well be a physical phenomenon, but he is also something more. This something more is a function of the manner in which men consider their relations to the community to which they belong, the social and moral domain. This aspect of human activity is covered in the manifest image of man. And so the philosopher's problem is somehow to account for the elimination of the physical issues from the manifest image and at the same time to tie in the scientific image so as to provide a complete framework.

Whether or not Sellars is correct in claiming that a scientific picture of the universe cannot fully accommodate all we wish to say about ourselves, we can still appreciate his observation that the scientific image of the world conflicts with that image based on merely a sophisticated common sense analysis of the phenomena we perceive. Humean efforts to explain empirical generalizations by deducing them from theories can be viewed as one attempt to resolve this conflict. Coupled with this effort is the assumption that whatever theory we use, the manifest image retains both its empirical and logical priority.

Unfortunately, the Humeans are caught on the horns of a dilemma. On the one had, if they insist both that scientific theories do all that is required of them and consider the reconstructed product of scientific reasoning as a deductive systematization of the data, they run headlong into Hempel's Theoretician's Dilemma and the conclusion that theories are eliminable. [16] If theories are eliminable then the possibility of generalized scientific knowledge evaporates.

On the other hand, if scientific theories are to include an *inductive* systematization of data, Hempel's solution to the theoretician's dilemma, then because of the notorious difficulty, if not impossibility, in using statistical laws to explain individual events and, hence, to provide inductive explanations, theories are again eliminable. They are designed to explain phenomena; if theories fail to provide explanations, then they fail to serve their purpose and there is no reason to appeal to them.

The Humean dilemma rests on the key methodological assumption that the manifest image has a kind of logical priority. Attempts to define the theoretical concepts of theories in terms of observational concepts is an essential aspect of the commitment to the manifest image. Sellars seeks to dissolve the dilemma. This requires two things. First, there must be a shift in our ontological commitments from the objects of manifest to those of the scientific image. Secondly, we have to transcend the dualism created by viewing these images as competing. This is accomplished by incorporating the facts of science into that conceptual scheme most appropriate to expressing what it is to be a person.

The scientific image results from taking the postulational approach to explaining the observable phenomena seriously. While Sellars recognizes the methodological hold the manifest image has on our thought, and that the scientific image in some sense is, moreover, an outgrowth of the manifest, he nevertheless believes that the scientific image actually constitutes a rival to the manifest image. The competition concerns the types of things in the world. His basic argument rests on characterizing each of these images in a manner which transcends individuals and thereby takes on a kind of objectivity. As such they are criticizable and even open to the possibility of rejection as false. ([45], [46])

I do not intend to explore all of the ramifications of Sellars' analysis of the objectivity of images here. Intuitively, the account has a certain plausibility and that is sufficient. Instead of taking on the larger issue now, I wish to merely sketch the kinds of considerations Sellars has in mind when he argues to the conclusion that the manifest and scientific images are conflicting explanatory frameworks and, moreover, that we can conceive of a possible resolution of this conflict. The problem of the objectivity of images emerges shortly, when the problems of justification are discussed. If science uses only those techniques of analysis and systematization which characterize the manifest image, that is, those methods Sellars lumps together under the heading 'correlational induction', then we cannot possibly explain all the unexplained phenomena, for the best we can do with these methods is to produce empirical generalizations which stand in need of explanation themselves. In those cases where the data only admits of a statistical generalization, then we cannot even explain individual cases.[6] Explanation entails telling what the objects or events *are*. To claim that there is a 70% chance that the thing in question is an X hardly counts as an explanation.

On the other hand, by postulating imperceptible entities to account for the behavior of perceptibles we *can* produce the explanations we seek. We can explain the single case in all instances by demonstrating that the perceptible is really a collection of imperceptibles, etc. We can explain the generalization by explaining what the perceptibles whose behavior they describe really are and, thus, why it is that they behave in the way the generalizations, universal and statistical, describe. This, of course, leads us to the point where we are forced to decide which set of objects are the *real* constituents of the universe, the perceptibles or the imperceptibles. This marks the major difference between the two images, their basic constituents.

To claim mere explanatory power leads us to reject one framework for another oversimplifies the issue. At stake is the make-up of the universe in which we live. Are there really trees and birds out there or merely collections of energy pockets functioning within the structure of a field of some kind? It seems that the final argument in favor of Sellars' scientific image comes to the point that while the scientific image can explain the manifest, the manifest cannot answer all its own questions, no less even begin to question the scientific.

This is not to say there aren't any questions concerning the scientific image. There are several, and the major one is very simple: What is its justification? With respect to the manifest image, justification is a relatively simple matter. If the image entails something about man or the world, we look and see if it is indeed so. But how can we challenge a system which *begins* by postulating unobservables and rests a great deal of its case on the reality of those entities and denies the legitimacy of the grounds of the opposing framework? If the content of the system can't be challenged, then there is no question of justification. The justification of the system fades out of the picture as a real problem. We are simply left with what we have. But, as Sellars points out, criticism is an important aspect of scientific inquiry. ([45], p. 7) And so to be truly scientific this image of man must admit of criticism and to be acceptable must meet that criticism. To accomplish the latter a criterion of success is needed.

The Sellarsian response to the question of criticism is a double-hinged theory of justification based on the concept of explanatory coherence. First, it is important to realize that only in *reconstructing* the scientific image is it correct to say it *begins* by postulating unobservables. The scientific image is a genetic outgrowth of the manifest. So first we must distinguish the process of developing the scientific image from its justification.

The question of justification concerns the scientific image which is not a present fact. It is in the process of becoming. Thus, it is incorrect to speak of *the* scientific image at present. For currently there are as many images as there are independent sciences.

... the conception of *the* scientific or postulational image is an idealization in the sense that it is a conception of an integration of a manifold of images, each of which is the application to man of a framework of concepts which have a certain autonomy. For each scientific theory is, from the standpoint of methodology, a structure which is built at a different 'place' and by different procedures within the inter-subjectively accessible world of perceptible things. Thus 'the' scientific image is a construct from a number of images, each of which is *supported by* the manifest world. ([45], p. 20)

The final scientific image, what I will now call the final theory *FT*,[7] will be an explanatory system based on individually successful sciences coherently organized into one structure. The structure must not only be internally coherent, but because of Sellars' conviction that the scientific image cannot handle all the complex issues concerning *man* in the universe, there must be also a kind of external coherence at work with whatever framework is being developed to help us deal with problems other than those of physical explanation.

The internal coherence requirement also extends beyond the coherence of individual theories. Explanatory coherence is the condition of adequacy for individual theories as well as for *FT*. ([44], p. 122) With respect to the concept of explanation at work here, Sellars does not agree with Hempel that explanation necessarily entails subsumption of the description in question under a law. For Sellars, to explain an object entails telling what the object is. The very plausibility of the postulational method of science rests on this idea.

3. PICTURES AND COHERENT IMAGES

The notion of coherence Sellars uses involves the concept 'picturing'; a coherent image generates a coherent set of pictures. Not only is the concept of picturing essential to understanding the notion of coherence, it is also central to Sellars' account of induction. A picture is a linguistic item intimately tied to the concepts of a matter-of-fact and truth.

... to understand the point of inductive reasoning one must understand the distinctive functions of matter-of-factual statements belonging to the level below that of law-like statements ([48], p. 118)

Concerning matter-of-factual statements, two initial points should be made. First, there are two types of matter-of-factual statements: atomic and molecular. Both types pick out pictures, but "it is atomic statements which make up 'linguistic pictures' of the world". In their case "the subject term is a definite description". ([48], pp. 119; 124) In the case of molecular statement, "they pick out sets of pictures within which they play no favourites ... " ([48], p. 119) That is to say molecular statements of fact are general descriptions in which it would be inappropriate, if not nonsensical, to speak of definite descriptions occurring.

The second, and perhaps more important point to be made concerning matter-of-factuals is that the vocabulary in which it is appropriate to phrase atomic matter-of-factual statements is not limited to the terms of a Humean object language. More precisely, singular sentences using theoretical terms also provide significant pictures.

Picturing is a relational linguistic activity. It is not the activity of asserting matter-of-factual propositions. Rather, it is the world in linguistic form. That is a peculiar claim, but consider Sellars' characterization:

Linguistic picture-making is not the performance of asserting matter-of-factual propositions. The *criterion* of the correctness of the performance of asserting a basic matter-of-factual proposition is the correctness of the proposition *qua* picture, i.e., the fact that it coincides with the picture of the world-cum-language would generate in accordance with the uniformities controlled by the semantical rules of the language. Thus the *correctness* of the picture is not defined in terms of the *correctness* of a performance but vice versa. ([48], p. 136)

Thus picturing is not asserting a description of a matter-of-fact, it is the matter of fact. The correctness of asserting the fact is judged by comparison with the picture. That is to say that the correctness of an assertion is determined by the semantical rules governing asserting. These rules determine when an assertion is possible, and the verification of the decision resulting in the assertion is the picture itself. It is in this sense that we see the import of Sellars' claim that "the *observational application* of a concept cannot be the obeying of a rule at all. It is essentially the actualization of a thing-word $S-R$ connection". ([47], p. 334)

A picture is a linguistic item. It is a singular matter-of-factual statement, the subject term of which is a definite description. The subject term refers, not by pointing to objects, but by being correlated with them. That is, a statement is a linguistic picture uttered in the presence of the object concerning which the subject term of the statement is a definite description.

The fundamental job of singular first-level matter-of-factual statements is to picture, and hence the fundamental job of referring expressions is to be correlated *as simple linguistic objects* by matter-of-factual relations with single nonlinguistic objects. The difference between '*a*', on the one hand, and 'the *g*' and '$(\iota x)gx$', on the other, is that the latter carry on their sleeve the logical and empirical information relevant to their correct use. ([48], p. 124)

Sellars calls for a *correlation* between linguistic and non-linguistic items. The correctness of the assertion is a function of the correctness of the correlation. This requires an agent who knows how to correctly correlate the items. To understand the consequences of this view, we need to look more closely at the distinction between 'referring expression' and 'refers'.

Names do not simply refer; people refer. Names and other expressions are *used* as referring expressions when rules are available for specifying the conditions under which the phrase can be correctly, as opposed to incorrectly, correlated with a non-linguistic item. To say 'correctly applied' entails that there are rules available concerning the correlation. The import of the Sellars' concluding claim that expressions like '$(\iota x)gx$' carry the logical and empirical information essential to their proper use on their sleeve extends beyond the now familiar analysis of meaning in terms of use.

To know the difference between 'a' and '$(\iota x)gx$' *is* to know their different uses. Rules govern the use of language, so to know the meaning of a word or phrase is to know the rules governing its use. But this is not as simple a view as it first appears to be. Knowing the difference between referring and other expressions also entails knowing what it is to be a referring expression. To be a referring expression is to be capable of being correlated with a non-linguistic object. But not just any object will count. We first need to know what are the referring expressions. This information is contained in the definite descriptiveness of those terms. Once we know what it is to be a referring expression and how to pick one out, learning the rules governing the correct application of the rules for each of the referring expressions in the language constitutes the final problem. This may sound like a large job, but Sellars is right in that it is part of what understanding a language comes to. Exactly how this is to be accomplished is yet another question.

The sense in which it is only atomic matter-of-factual statements which make up linguistic pictures at least should be clearer, if not opaque, at this point. To say of the subject term of a generalization that it is a definite description would result in the generalization picturing a class of entities. This is both a strange sense of referring and a stranger sense of matter-of-factual statements. Matters-of-fact are, for Sellars, true statements. The objects to which they refer must, in this context, be *in* the world. For a generalization to refer classes would have to exist in the world (however else they may be thought of to exist at this moment is irrelevant). And while I can contenance the idea that classes are abstract entities and have a reality of a kind, they are certainly not *in* the world like tables, chairs and doors.

Atomic sentences using both observational and theoretical terms can play the picturing role. This suggests that theoretical claims can be used as instruments: instruments to construct pictures of the world. ([48], p. 143) And while Sellars recognizes that this is a viable possibility, he is not anxious to be labeled an instrumentalist.

The Positivist understanding of the relation between science and our general common sense framework seeks to reduce the scientific image to the common sense image. Now if we also take into account the dilemma ([16]), wherein positivists are forced to conclude that theories are eliminable, it would seem that there is only one option left open to them. When questioned as to the value of the product of science, they must consider scientific theories only as useful instruments for facilitating measurements and observations. Sellars appears to consider only the instrumentalist as a serious opponent on the question of the priority of scientific over the manifest image. And while he never offers a reduction of opposing views to instrumentalism, something like it must be behind the key role arguments against instrumentalists have in his writings on scientific method.

The conflict between Sellars and such instrumentalists as Nagel centers on the picturing role of theoretical atomic matter-of-factual statements.

The instrumentalist, from our point of view, is one who holds that theoretical statements of all kinds, including singular statements, are *essentially* instruments for generating statements *in the observation framework*. Thus, if he went along with our distinctions he would hold that (ampliative) theoretical statements are simply more sophisticated instruments which along with molecular, quantified, and law-like statements in the observation framework are means of constructing *observation framework* pictures of objects and events. Picturing, to put it bluntly, would be the inalienable prerogative of the perceptual level of our current conceptual structure. ([48], p. 144)

But such a position entails a final ontological commitment to the objects of the manifest image. And because of this commitment, instrumentalists fail to see the possibility of the pictorial singificance of theoretical singular statements and, hence, a good deal of their cognitive singificance as well.

Instrumentalists (and philosophers of science generally) lay little stress on the role of singular statements in microphysical theories. They concentrate, rather, on the relation of theoretical principles to empirical laws; and the singular statements they emphasize are observation framework statements. Again, for the most part, they do not explicitly recognize the picturing dimension of factual truth, or fail to distinguish it clearly from the '*p*' *is true* ↔ *p* dimension. Thus even when they recognize the existence of properly singular statements in theoretical discourse, and recognize that the latter are properly characterized as true or false, this does not raise for them the question whether these statements can be regarded as conceptual pictures in their own right. ([48], p. 144)

To refer briefly back to the criterion of adequacy for theories, explanatory coherence, the kind of coherence Sellars has in mind can now be understood. He requires that an adequate theory produce coherent pictures which taken together constitute a clear image of the world. The items that must cohere are (a) theoretical atomic statements, (b) non-theoretical atomic statements, (c) molecular theoretical statements, and (d) quantified and law-like statements.

4. TRUTH AND EXPLANATION

The criterion of adequacy for pictures involves the concept of S-assertibility. And because S-assertibility and truth are intimately related in Sellars' analysis we can expect they will receive similar treatment, which they do. Sellars defines 'truth' as S-assertible.

> ... for a proposition to be true is for it to be assertible, where this means not *capable* of being asserted (which it must be to be a proposition at all) but *correctly* assertible; assertible, that is, in accordance with the relevant semantical rules, and on the basis of such additional, though unspecified, information as these rules may require .... 'True', then, means *semantically* assertible ('S-assertible') .... ([48], p. 101)

For a statement to be S-assertible is for that statement to meet all the requirements embodied in the relveant semantical rules governing the circumstances under which it is proper to make the statment, i.e., within some conceptual scheme, CS_i.

Now, for a picture to be a good picture it too must be S-assertible.

> ... pictures, like maps, can be more or less adequate. The adequacy concerns the 'method of projection'. A picture (candidate) subject to the rules of a given method of projection (conceptual framework), which is a correct picture (successful candidate), is S-assertible with respect to that method of projection. Thus the S-assertibility of a matter-of-factual proposition formulated by the schema
>
> The · fa · is S-assertible *quoad* CS_i
>
> is a matter of · fa · s being elements of correct pictures of the world in accordance with the semantic rules of CS_i. ([48], p. 135)

A good picture is a well-projected picture. A well-projected picture is one that is made in accordance with the rules of the given conceptual system in which it is formulated and the occurrence of the type of entity mentioned in the picture as a legitimate object in the system, and correctly matches the proposition with the occurrence of the object named by fa in the world.

Thus, an adequate picture is one made in accordance with both the semantic and pragmatic rules of a language. An adequate theory is one which produces a set of coherent pictures, all of which are assumed to be good (for no one would want a coherent set of bad pictures).

Now that we have an account of picturing we also have an analysis of one half of the criterion for a good theory. A good theory must meet the criterion of explanatory coherence. For a theory to cohere it must produce a coherent picture of the world. For a theory to explain it must tell us what the objects *are* pictured in the image of the world it produces. This is not a requirement that reduces to merely naming the objects. It is, rather, a demand for a statement of basic natures. Granted, at some point the quest must stop. But the criterion for an explanation is that it tell what the objects are, not the objects of the world, but the objects as already named and occurring in the linguistic picture. This is done by redefining the concepts in theoretical or imperceptible-item terms. As we shall see, the concept of a substantive correspondence rule, which serves the function of redefining observables in theoretical terms, plays a major role in Sellars' analysis of the structure of a theory. In his use of correspondence rules Sellars follows Aristotle's dictum that "A 'definition' is a phrase signifying a thing's essence". (*Topics*, 101b39)

5. EXPLANATIONISM

In so far as explanatory coherence is the criterion of adequacy for a theory, Sellars is an explanationist. Keith Lehrer describes 'explanationism' in the following way.

A belief is justified by its explanatory role in a system of beliefs. Some beliefs are justified because of what they explain, and other beliefs are justified because they are explained, but every belief that is justified is so either because of what it explains or because of what explains it. These doctrines formulate a theory of justification which I shall label 'explanationism'. ([22], p. 100)

Explanationism, as descibed by Lehrer, differs slightly from Sellars' theory of justification. First, Sellars justifies theories, rules of inference, empirical generalizations and descriptions instead of beliefs. Second, not only are these items justified because of what they explain, but some items are justified because they enable us to offer explanations.

In Sellars' version of explanationism we are justified in accepting a theory if it *explains*, a description if it *is explained*, and a rule of inference if it *helps to explain*. To explicate this view we need, therefore, an analysis of those items which this theory of justification uses, e.g., theories, rules of inference, and descriptions, as well as an analysis of 'explains'. The full import of the Sellarsian view of the role and nature of both theories and rules of inference can only be understood in terms of his theory of justification. Without his analysis of when and how we are justified in accepting theories and rules of inference a number of his claims would be at best idiosyncratic. With it, however, the system makes a great deal of sense, and is worthy of further investigation and development.

But just as his analysis of a theory without his discussion of justification fails to stand on its own, his theory of justification requires the rest of his analysis of science. Consider the following brief outline. According to Sellars we are justified in accepting theories only when they constitute explanatory frameworks. To have an explanatory framework entails being able to draw inferences about the phenomena contained in the domain of the theory. To legitimately draw inferences requires rules of inference. The occurrence of specific rules of inference in a theory is justified when they can be shown to contribute to the explanatory power of the theory.[8]

In this view we justify the *use of rules* of inference with respect to explanations. This differs from *DN*[9], where we justify the *laws* used in explaining a given claim. For Sellars a theory has two parts: a formal structure, viewed as an object language, and a meta-language containing rules for manipulating the object language. Problems of justification pertain to both the constituents of the meta-language and the theory as a whole. According to Sellars, analyzing a theory as only an interpreted abstract formal language severs the crucial relation between the logical structure and the users of the structure. Understanding the nature of theories requires incorporating into your analysis the fact that theories are designed for specific uses. A theory is a device for reasoning about the world, not merely an interpreted formal system. Sellars stresses this point by embedding the logical structure of a theory in a wider context of practical reasoning.

But recognition of this aspect of theories is not Sellars' private insight. Even though Hempel begins by attempting a purely syntactic reconstruction of a theory, he moves step by step to the point where finally, [14], he too takes into account the same point about the essential role of the users of the theory, at least for justificatory purposes. However, Hempel introduces the notion of epistemic utilities to handle the problem while Sellars uses the concept of practical reasoning.

Because theories are used to give explanations, problems of justification pertain primarily to the meta-language of a theory. Explanations are produced by deductively drawing inferences about entities described in the object language in accordance with the rules of inference of the meta-language. If the result fails to yield an explanation, or if a prediction based on the rule fails to turn out, the rule must be rejected. The acceptance of new rules must then be justified. And, according to Sellars, this issue differs from the formulation of empirical generalizations, located in the object level, on the basis of evidence of a given kind. The problem here is justification of the rules, located in the meta level, which permit both the formulation and use of those generalizations.

Sellars is pushed to an explanationist position on the issue of justification. Because he believes that scientific inquiry has a purpose, a goal toward which it strives and, moreover, because of his insistence on the use of deductive reasoning when working within a theory, the basic Humean approach is closed to Sellars. He acknowledges what so many Humeans continue to resist, that empirical generalizations cannot be justified if they are the product of induction over observation formulated in the framework of the manifest image. The goal of science is to replace the manifest image with a more refined one, hence it would be somewhat strange to expect to justify the replacing image in terms of the replaced image. On the other hand, if the logic of scientific theories is deductive, the cash value of theories would appear to rest in the conclusions drawn from them. For Humeans who judge a theory in terms of its observational results, this is fine.

For Kantians, taking the position that observations are 'theory-laden' in such a way that they are insignificant apart from the theory under which they are made, the deductive nature of theories merely serves as a means for checking consistency. And while consistency may be part of a criterion of adequacy for a theory, it surely is not the whole story.

Sellars, while neither a pure Humean nor a pure Kantian, produces a methodological synthesis which does justice to the insights of both approaches. He shows his Humean colors by emphasizing the manner in which the scientific image replaces the manifest: by redefining the objects of the manifest image, so as to better understand them. He does not merely dismiss it. Thus, the manifest image retains its importance as a touchstone. But, in the Kantian spirit, Sellars gives the final word to the scientific image and the theories it uses to deductively explain the phenomena of the perceptible world. To effect this synthesis, Sellars distinguishes the product from the process of science, but unlike Popper, does not then dismiss the process. As we have seen, he believes that it would be both methodologically misleading and historically inadequate to do so. But, faced with the questions of justification, the decision must be made with respect to what we do or what we produce. That is, if we distinguish between theories and the manner in which they develop, to which side do we direct a theory of justification? Do we justify the end product of our efforts? The answer, of course, is to justify both. Let us briefly consider some of the issues involved.

(1) We justify the *process* of inquiry by first characterizing the process in terms of the policies we use in constructing theories. We then justify the policies by showing they lead to the final theory which is the goal of science

and for which purpose they were adopted in the first place. It is at this point
that the teleological dimension of Sellars' view can be discussed.

Sellars is a teleologist to the extent that he believes science is concerned
with constructing a complete coherent theory for the purpose of replacing
the explanatory framework of the manifest image. Moreover, the progress of
science is to be measured in terms of approaching this goal. He claims induction is a rational mode of reasoning. In this way he can legitimize the activities
of the framework in which theoretical entities are postulated and inductive
generalizations are formulated.

Thus, it is rational to postulate new entities if one is doing science, i.e.,
engaged in the attempt to construct theories. But as we have seen, to be
engaged in science involves a commitment to the reality of the scientific
image. This, furthermore, can only be justified in the long run when FT is
complete. This long range justifiability is what the criterion of explanatory
coherence in conjunction with Sellars' appeal for a stereoscopic vision of man
in the universe ends up involving. The scientific image will be unacceptable
unless it can be shown to be coherent with the concept of a person, i.e., unless it can be used.

But, it might be objected, does this not make Sellars a pragmatist rather
than a teleologist? In fact, Sellars seems to believe that he is a pragmatist since
he adopts the pragmatist principle that knowledge entails action, amending
it to read 'knowledge of a language'. ([47], p. 340) Furthermore, his concern
with practical reasoning as the basic form of reasoning leading to action
appears to substantiate this claim. However, claiming that knowledge and
action are related and that practical reasoning is all-pervasive does not make
one a pragmatist.

That part of pragmatism which Sellars invokes concerns the relation of
language to action. To be able to make a decision to act entails knowing a
language in which the reasoning culminating in the decision can be formulated. ([47], p. 340) Language is the logical prerequisite for rational action.
The purpose of language is to make rational action possible. But despite
Sellars' concern with the explanation of language in terms of its significance
for our understanding of rational action, he stops short of full scale Peircean
pragmatism. Sellars claims "it can be reasonable to *accept* a hypothesis which
is not reasonable to *act on*". ([42], p. 228) Peirce says, "Belief does not
make us act at once, but puts us into such a condition that we shall behave
in some certain way, when the occasion arises". ([25], p. 10) The key to
understanding the difference between Sellars and Peirce here rests on the
'when the occasion arises'. There may very well be situations in which several

things you believe bear on what you ought to do. But that you fail to act on some of these beliefs does not entail that you ought never to act on them or, Sellars' point, that it is *reasonable* not to act on them. The manner in which Sellars phrases the entire issue suggests some confusion. For example, if he means it is not reasonable to act on something you accept as true and hence, believe, does he mean it is conceivable that there are things you believe on which you would never act? Or, does this position entail something weaker? I believe Sellars argues the stronger case and hence renders the notion of acceptance unintelligible.

But if Sellars is not a pragmatist, he *is* a teleologist in the sense that Peirce was. Peirce believed that belief was a habit to act in a certain way. One of the things that belief considered in its perview was what was real. What you believed real was real, or what amounts to the same thing: what you were inclined to treat as real in your actions constitutes reality. But this did not entail solipsism since the final determination of reality was a community decision.

... reality is independent, not necessarily of thought in general, but only of what you or I or any finite number of man may think about it; and that, on the other hand, though the object of the final opinion depends on what that opinion is, yet what that opinion is does not depend on what you or I or any man thinks. ([28], p. 39)

Peirce claims here that reality is a function of the continual interchange of ideas among men. It is dependent upon thought, but not limited to the thought of one man, or limited group. To call it independent of one man and yet dependent on thought can only mean that what is real is what we take to be real as it is embodied in our language or conceptual scheme. That those items change is part of what is entailed by the allusion to the independence of reality from the beliefs of a finite number of men. What is real is what all men believe. Moreover, there is a means by which we can find that reality: the method of science. This method is "such that the ultimate conclusions of every man shall be the same". ([25], p. 18) Since "the irritation of doubt causes a struggle to attain a state of belief", ([25], p. 18) and that which all men come to agree on is what is real, then we must simply come to a point where the doubt of all men as to what is real is eliminated.

Our perversity and that of others may indefinitely postpone the settlement of opinion; it might even conceivably cause an arbitrary proposition to be universally accepted as long as the human race should last . . . . But the reality of that which is real does depend on the real fact that investigation is destined to lead, at last, if continued long enough, to a belief in it. ([23], p. 39)

As Peirce notes, we may get side tracked, but our final agreement, in such a way to assuage all doubt, will constitute the fact of reality. Peirce, therefore, considers inquiry and its culmination in a final scheme of belief to be the job of science, in much the same spirit that Sellars acts.

There is, however, a crucial difference between the two men on this point as well. For while Peirce does not claim that it is the purpose of doubt to lead us to belief, Sellars does claim that the purpose of postulating imperceptibles is to arrive at a final explanation which replaces that of the manifest image. Both agree that inquiry will end with a complete system. Moreover, both agree that the purpose of science is to settle doubts and provide explanations. But Sellars takes a step beyond when he speaks of the postulational method of science. For clearly such a break with what we might call common sense would require some justification. Peirce, on the other hand, does not find himself in the quandry of having to justify doubt.

Sellars' justification for the process of science, the postulation of imperceptibles, the critical analyses of hypotheses and rules of inference, etc., is the construction of *FT*. We are justified in postulating imperceptibles if we can construct a final theory which meets the conditions of explanatory coherence. This then leaves us with the problem of justifying the product of science.

(2) We justify the *product* of science by showing that it does the job it was constructed to do: provide explanations. At this point Sellars' deductivism enters. An explanation tells us what the object or item in question is. This is accomplished either by redefining the item in question in terms of what is believed to really be the case or by showing there is a connection between the item to be explained and what is believed to be the case. In the first case we explain by definition. The question remains concerning the type of connection called for in the second case.

Sellars argues that there is no such thing as inductive arguments, hence there can be no inductive or non-demonstrative explanations. ([42], pp. 201–202 and [39]) The crux of his argument lies in his account of logic.

Without attempting to define what is meant by a 'logic', it seems reasonable to say that, however many 'logics' there are, they are 'logics' by virtue of their concern with what makes an argument sound. In the case of 'deductive logic', the concept of a sound argument is that of an argument which is such that if its premises are true, its conclusion *must* be true. ([39], p. 83)

Since we cannot come up with rules which guarantee soundness for non-demonstrative arguments, there are no non-demonstrative logics. Science tells what really is the case. And since the heart of explanation is to relate what we seek to understand to what we know and since the only *systematic* way of establishing that kind of relation is by using the rules of deductive inference, deduction must be the logic of science.[10]

Using scientific theories we reason deductively to explanations. The sense in which deductive arguments constitute the only legitimate systemization of data is a major Sellarsian theme. The idea is closely tied to his analysis of intentions and moral reasoning. These various topics are all tied together under the rubric of a general theory of practical reasoning. A relatively simple theory, it has only one major principle:

As for the logic of intentions, i.e., of practical reasoning, it is governed by one simple principle: if

if (. . .) implies (- - -)

then

[shall (. . .)] implies [shall (- - -)] . ([42], p. 205)

The theory becomes more complex when this principle is applied to the fruits of theoretical reasoning and empirical investigations. But the sense in which this principle bears on Sellars' deductivism turns on how practical reasoning supposedly lies behind the use of theories. He distinguishes between the approariate reasoning for using theories and that type of reasoning which leads to the formulation of generalizations and the postulation of theoretical entities.

Inductive reasoning is the *mode* appropriate to the construction of theories. But this does not entail that there is an inductive *logic*. The distinction between the inductive and deductive modes of reasoning and the acceptance of inductive reasoning as rational, without further commitment to an inductive logic, marks a major aspect of Sellars' theory of science. Sellars' point here, one we consider in greater detail in the following chapters, is that the method of science, i.e., the method by which it progresses, is inductive. Moreover, it must be inductive if science is to be a rational inquiry.

But with respect to justifying theories, according to Sellars we justify a theory by showing that we can indeed deductively argue from what we believe to be true to what we don't understand. In this way we establish a relation which is open to every man to investigate (in the spirit of Peirce).

In the remaining chapters I will try to show how Sellars' belief in the eventual success of science and in the absence of other than deductive connections leads him to justify both *FT* and the process which led to its construction in explanationist terms. The stage was set by Sellars' initial characterization of the scientific method as postulational; it remains to see how the characters play out their parts.

CHAPTER II

RULES OF INFERENCE, INDUCTION, AND AMPLIATIVE FRAMEWORKS

1. AMPLIATIVE INFERENCE

Sellars distinguishes the reasoning appropriate to *developing* hypotheses and theories, i.e., induction, from that appropriate to *testing* and *justifying* its results, deduction. Induction is a legitimate mode of inference; it must be if scientific inquiry is to have any respectability. Sellars not only argues that induction is legitimate, but that empirical science in general is essentially inductive. ([47], p. 355; [40], p. 304) Empirical science proceeds within an inductive framework: a framework which accepts inductive inference as rational.

Now while the general framework of science is inductive, reasoning within a theory is deductive. Deductive reasoning is characterized by rules of inference which delineate sets of possible inferences. Rules of inference are, of course, rules for making inferences. They are important because, as I would also argue, they provide the regulative basis for making judgments and reasoning about the world. 'Regulative' in this context should be distinguished from 'constitutive'. A rule operates *constitutively* when it is the *a priori* basis for asserting the existence of something. To talk about the existence or non-existence of some entity the concept used to name the entity in question must be available. For example, to identify an object as red the observer must have the concept of a red thing at his command.

A rule is used *regulatively* if it applies to or merely determines the relations between things already accepted as existing. Thus, rules of inference permit judgments because they delimit the class of possible relations between entities whose ontological status is already determined.[1]

Rules of inference are used within theories, systems, or language. These in turn are functionally characterized in terms of a set of rules operating at a meta-level, rules concerning the nature of the theory, etc. The meta-rule governing the enterprise characteristic of the theory is analogous to a constitutive rule, insofar as it licenses the activity. To label empirical science 'inductive' requires that induction be a constitutive dimension of empirical science. This claim is common to Sellars, Goodman, and Quine. The reasoning that accompanies such a view of the foundation of scientific thinking varies, however, with respect to the analysis of the constitutive nature of inductive inference. For Sellars it remains a logically necessary condition for the development of science. Goodman claims it is a fact with which we must learn to live. ([12], p. 64) Quine resuscitates his naturalism by discussing induction in terms of its survival value. ([32], p. 125)

But merely calling induction a constitutive dimension of empirical science invites confusion. Let us distinguish with Salmon between 'ampliative' and 'inductive'. Using 'ampliative' as the name of the general category of inference types, Salmon says that an ampliative inference "has a conclusion with content not present either explicitly or implicitly in the premises". ([36], p. 142) He also distinguishes between ampliative and non-ampliative, where a non-ampliative inference is deductive. Inductive inferences form a subset of ampliative inferences. The set of ampliative inferences includes all inferences in which the conclusion is not contained in the premises. In inductive inferences we infer from *true* premises with insufficient evidence to claim generality to what we hope to be a *true* conclusion. The old traditional problem of inductive inference has been to find rules which will guarantee that the conclusion will be true, given insufficient evidence for generalizing the premises (and turning the argument into deductive form).

What Sellars means, or what he should be saying when he calls induction constitutive of empirical science, is that ampliative inference is a legitimate aspect of scientific reasoning. We must be able to construct general hypotheses and theoretical principles if we are to build a theory on scanty information. In this context ampliative inferences are inferences to something unknown on the basis of what we already know. In this sense of 'ampliative' science is constitutively ampliative. But 'science' here means the inferential process of inquiry and not just some product or conclusion of an inference. The scientific process of seeking out the truth must be ampliative in nature if we are to go beyond sense experience.

Both Quine and Goodman confuse making particular inductive inferences with the nature of ampliative inference in general. Goodman correctly argues that the problem of induction is not to justify induction in general but to come up with a way of justifying particular inductions. Induction in general, a logically necessary condition for a developing science, cannot be empirically justified. Thus, in the framework of a developing science it is rational to make ampliative inferences.

Quine's account is at best unhelpful. He rightly notes that making inductive inferences is a 'natural' thing do. Unfortunately, his use of this insight leads to a great deal of trouble.

Of course, these unadorned attempts to portray accurately a crucial aspect of these philosophical positions easily lend themselves to charges of being unsympathetic or patently false. So let us look at the views of Goodman, Quine, and Sellars on the constitutive nature of inductive (ampliative) reasoning and see why Goodman and Quine fail to make their case.

2. SELLARSIAN RULES OF INFERENCE

Towards the end of "Some Reflections on Language Games" Sellars remarks that "an understanding of the role of material moves in the working of a language is the key to the rationale of scientific method". ([47], p. 355) In his description of a language game, a material move is an inference from one object level statement to another, the validity of which depends on the predicates contained in the two statements. Thus:

(1)   There is smoke

to

(2)   There is fire

is a material move. We sanction the enthymematic inference from (1) to (2) when asserted in appropriate circumstances. The meaning of (2) relates to the meaning of (1) in that fires sometimes produce smoke. A material move is a valid inference of this kind.

On the other hand, the inference from

(3)   There is smoke

to

(4)   There is water

is not valid. The type of connection that exists between 'smoke' and 'fire' does not hold for 'smoke' and 'water'. In other words, for '(1) to (2)' to qualify as a material move requires (a) a non-analytic relation between the meanings of the terms, and (b) a rule which permits the inference.

The rule which says we may infer (2) from (1) is an interesting creature. It is closely related to, but should not be confused with, a law of nature. Laws of nature as formulas for generating descriptions occur in an object language, while rules of inference occur in a meta-language which parallels the object language. Sellars sees the relation between rules of inference and laws in the following way: "to say that it is a law of nature that all A is B is, in effect, to say that we may infer 'x is B' from 'x is A' ". ([47], p. 331) But, "if I learn that in a certain language I may make a material move from 'x is C' to 'x is D', I do not properly conclude that all C is D". ([47], p. 331) Thus all laws, on this view, are rules of inference, but not all rules of inference are laws.

Now insofar as the move from 'all observed x's are C' to 'all x is C' constitutes a typical example of inductive inference, two possibilities present themselves as explanations of the claim that not all material moves entail laws: (a) induction is not permissible, and (b) knowing a given material move is possible does not entail knowing the rule or a law of nature.

Since scientific reasoning is constitutively inductive, (a) seems unlikely. To understand why (b) is the proper alternative requires both Sellars' distinction between 'rule-obeying' and 'pattern-governed' behavior and a discussion of rationality in science.

Rule obeying behaviour contains, in some sense, both a game and a meta-game, the latter being the game in which belong the rules obeyed in playing the former game as a piece of rule obeying behaviour. ([47], p. 327)

The rules of a game are not part of the game, as a roll of the dice would be. They are about the game, as in: 'one must roll the dice to find out how many spaces to move'.

Pattern governed behavior is best described in terms of how Sellars sees it learned.

To learn pattern governed behaviour is to become conditioned to arrange perceptible elements into patterns and to form these, in turn, into more complex patterns and sequences of patterns. Presumably, such learning is capable of explanation in $S-R$ reinforcement terms, the organism coming to respond to patterns as wholes through being (among other things) rewarded when it completes gappy instances of these patterns. Pattern governed behaviour of the kind we should call 'linguistic' involves 'positions' and 'moves' of the sort that *would be* specified by 'formation' and 'transformation' rules in its meta-game if it *were* rule obeying behaviour. Thus, learning to 'infer', where this is purely a pattern governed phenomenon, would be a matter of learning to respond to a pattern of one kind by forming another pattern related to it in one of the characteristic ways specified (at the level of the rule obeying use of language) by a 'transformation rule' – that is, a formally stated rule of inference. ([47], p. 327)

Thus, it would appear, one can infer uncritically. That is, one can make inferences, given data (linguistic and experiential) and having learned certain responses via $S-R$ techniques, without knowing that a general descriptive claim regulatively governing talk of A's and B's also holds. Moreover, one can do this because of the structure and rules of the language in which he phrases his responses and inferences. Or, as Sellars puts it,

we may say that it is by virtue of its *material* moves ... that a language embodies a consciousness of the lawfulness of things. ([47], p. 331)

While the language allows drawing certain conclusions by permitting given inferences, the statement of which can be viewed as instances of general laws, some particular user of the language may not know all this.

Material moves mark off an 'awareness' of lawfulness, but it is, properly put, incorrect to equate the ability to make material moves with inferring. In the same sense in which someone who makes the correct moves but does not know the rules of a game cannot be said to fully know how to play, the mere making of material moves does not constitute inferring,

... unless the subject not only conforms to but obeys syntactical rules (though he may conceive them to be rules justifying the transition from one *thought* to another, rather than from one linguistic expression to another), so that he is able to *criticize* verbal sentences. ([47], p. 334)

But, from insisting the subject know there are both rules and moves in accordance with rules if we are to describe him as *inferring*, it does not follow that he need know *every* rule operative in a language. To be able to infer, according to Sellars, requires knowing a language. We describe the language by delineating its rules of inference, syntax and vocabulary. On the other hand, knowing a language does not entail knowing every rule. Some inferences are made for which the governing rule is not known. This entails that something other than rules of inference, syntax and vocabulary is also operative.

We can unearth this 'something other' by going back to the idea that the subject must be able to *criticize* verbal sentences if we are properly to describe him as inferring. But in this context, 'criticism' is ambiguous. The subject can criticize verbal inferences for at least wo possible reasons: either the conclusion-sentence is grammatically incorrect or the inference is incorrect.

Sellars does not elaborate on the notion of criticism and hence fails to make clear what else is at stake here. The ambiguity noted above permeates the rest of his discussion. In what follows I attempt to resolve the ambiguity in favor of the second alternative noted. In the context of inductive reasoning, the subject must not only criticize the sentence as the conclusion of an inference, but his criticism should not rest primarily on syntactical considerations. The issue is a pragmatic one, but the pragmatism is bracketed and, to continue the metaphor, the quantifier is ontological.[2]

To say 'Jones infers' entails that Jones both conforms to and obeys syntactical rules. To say 'Jones conforms to syntactical rules' means we can describe Jones' linguistic behavior in terms of a set of rules, all the members of which *he* may not know. If Jones obeys syntactical rules, then Jones' linguistic behavior is intentionally governed by reference to what those rules say is permissible and what is not permissible. For brevity, call a verbal utterance a sentence. Now, if Jones criticizes a sentence as poorly formed (to take the first alternative into consideration) and if this is a mark of Jones' ability to infer, then since the criticism is grammatical, Jones must know the rule governing correct sentence formation. But when he cites this rule as a justification for his criticism, he gives a reason why the sentence is incorrect and he argues deductively. On the other hand, we can distinguish judgments from inferences if we consider the judgment 'That sentence is incorrect' as evidence of Jones' having made an inference. A judgment is an assertion.

To illustrate the constitutive nature of inductive reasoning we have to consider the concept of criticism when what is criticized is the inference to the conclusion. But this too must be further qualified. Not only must the inference to the conclusion be criticized, but it must be criticized in the absence of knowledge on Jones' part of a rule governing that move.

If there is no known rule, then what could be the grounds for Jones' criticism? Consider first the fact that given an unchallenged inference, i.e., move, we can construct a rule which could govern future moves of the same type. Knowing that a given material move is possible provides the ground for discovering a rule, provided Jones understands the full complexity of the structure of his language. In this case then he may not know the particular rule, but if he understands the nature of material moves, he can figure it out. To understand the role of material moves entails knowing they

... can be characterized both as constituting the concepts of the language and as providing for inferences, explanations, and reasons relating to statements formulated in terms of these concepts. ([47], p. 355)

In other words, understanding the role of material moves entails knowing how induction functions as a constitutive part of language. One must first know the meanings of the terms occurring in different statements. Given this, inferences from one statement to another are possible and explanations of 'There is fire' are formulable in terms of the rule governing the move.

This constitutes the heart of internal explanation or Sellarsian explanationism at work. An internal explanation is a statement of fact, phrased in the object language and justified by appeal to a rule located in the meta-language which establishes a material connection between the observed phenomenon, 'There is smoke', and its explanation 'There is fire'. 'There is fire' explains 'There is smoke', and is itself explained by the rule of inference. The rule is justified because it functions as an explainer, thus keeping to the explanationist principle of claiming a statement is justified if it explains or is explained.

But notice, we have here an account of how one description can be said to be an explanation of another, but not what constitutes the mode of criticism which captures the sense of inductive inference. Material moves are the trappings of inductive inference, not the reasons why being able to criticize sentences is a mark of Jones' ability to infer.

In the following passage Sellars implicitly recognizes the need to distinguish two senses of 'criticize'.

Everyone would admit that the notice of a language which enables one to state matters of fact but does not permit argument, explanation, in short *reason-giving*, in accordance with the principles of *formal logic*, is a chimera. It is essential to the understanding of scientific reasoning to realize that the notion of a language which enables one to state empirical matters of fact but contains no material moves is equally chimerical. The classical 'fiction' of an inductive leap which takes its point of departure from an observation base undefiled by any notion of how things hang together is not a fiction but an absurdity. The problem is not 'Is it reasonable to include material moves in our language?' but rather '*Which* material moves is it reasonable to include?'. ([47], p. 355)

Reason-giving constitutes the initial sense of criticism, paralleling our earlier discussion of the grammatical dimension of criticism. The second sense of criticism we can delineate is that necessitated by the material moves. If 'which material moves?' constitutes the key question, how do we answer it? A clue lies in the idea that, to rephrase Sellars' observations, so-called 'neutral' observation bases are defiled by prior notions as to how things hang together. The basis for criticizing one sentence, viewed as conclusion to an inference, is that it contradicts the image produced by the rest of the language and its material moves. Because of the implicit criterion of coherence embedded in this analysis, an incoherent move requires both that new material moves be tried and that rules governing them be formulated. If Jones obeys the rules of the game as well as conforms to them, he can criticize material moves and inferences because they fail to fit in with the rest of his world view. Unlike those cases where Jones cannot criticize a move because it might break a rule, where no particular rule exists he can criticize the move on the general grounds that it fits or fails to fit the rest of the image. That is, he can criticize on the grounds that it explains, fails to explain, or can be explained or not.

But this account of material moves does not by itself provide the understanding of scientific reasoning Sellars wants. For one thing, it is not clear why language should be described as a game, even though it seems to be a popular activity. Secondly, the outright acceptance of this view of induction is unreasonable, especially when the validity of inductive inference remains the source of continuing debate.

Casting about for some way to augment Sellars' account, it appears that when Sellars claims that the problem lies in deciding which material moves to include in our language, he moves in the same direction as Goodman. In *Fact, Fiction, and Forecast*, Goodman argues that Hume was correct when he said there was no justification for the move from

(1) $Crow_1$ is black
(2) $Crow_2$ is black
.
.
.
(k) $Crow_k$ is black

to

($k + 1$) All crows are black.

But, he also argues that Hume was on the right track when he said that we formed habits on the basis of past experience, which habits are the source of our belief that, for example, whenever we see a crow it will be black. Goodman sees this as Hume's attempt to define a valid prediction, i.e., one which arises from a habit which stems from experiencing past regularities. Reading Hume this way instead of in the traditional manner, wherein Hume argues only for the impossibility of justifying inductive conclusions, Goodman finds a clue for sorting out the confusions surrounding induction.

It dawns upon us that the traditional smug insistence upon a hard-and-fast line between justifying induction and describing ordinary inductive practice distorts the problem. And we owe belated apologies to Hume. For in dealing with the question how normally accepted inductive judgments are made, he was in fact dealing with the question of inductive validity. The validity of a prediction consisted for him in its arising from habit, and thus in its exemplifying some past regularity. His answer was incomplete and perhaps not entirely correct; but it was not beside the point. The problem of induction is not a problem of demonstration but a problem of defining the difference between valid and invalid predictions. ([12], p. 68)

On this reading of Hume, Goodman confronts the problem of defining the difference between valid and invalid prediction. Now insofar as Sellarsian material moves are valid inferences and the key Sellarsian question is 'Which inferences would we include in our language?', both Sellars and Goodman seek the same thing: a method of justifying sets of moves. Goodman locates the beginnings of such a method in his notion of entrenchment. But, the theory of entrenchment and projection is incomplete. It needs to be augmented by something like Sellars' account of a rule of inference.

In the ensuing discussion of entrenchment I consider some of the informal commentary Goodman adds to his proposed theory of projection and the extension Quine offers on the issue of artificial versus natural kinds. Goodman's ontological views, if read with the realist sympathy they suggest, are bewildering. (To forestall initial objections to calling Goodman a realist in any respect, recall that it is Goodman who envisages the theory of projection solving the problem of natural kinds. His proposed solution has realist overtones.) Quine's suggestion that in the end science eliminates the concept of kind is equally confusing. Quine's views are quite close to Sellars', and so after considering his ideas on natural kinds we return to Sellars' views on the development of science. The point here is that a full account of 'entrenchment' requires settling some things about the nature of science, in particular, the sense in which science can change. This also marks a good junction at which to compare the relative merits of these ideas. For the inadequacies of Goodman's account stem, I believe, from disagreement over what we ought to be worried about.

The methodological difference between Sellars and Goodman–Quine is captured by their divergent views on the objectives of science. Sellars sees science as striving toward the day when it can tell us what there is. Goodman and Quine see science as producing a body of knowledge to be used for solving problems.

Goodman believes that *at any given time* the product of scientific inquiry should be systematizable, hence the need to define 'valid inductive inference' to firm up the theory of confirmation and to find a means of distinguishing lawlike from non-lawlike empirical generalizations. Sellars, on the other hand, seeks an analysis of science that not only permits but justifies *scientific change*, a goal founded on the belief that to speak of science producing a product prior to understanding the context is misleading.

3. GOODMAN ON INDUCTION AND THE SCIENTIFIC FRAMEWORK

Goodman's renewed attack on the problem of induction begins, as we have seen, by suggesting that rather than seek a justification for induction, we should determine the difference between valid and invalid inductions. He identifies Hempel's work on the logic of confirmation as one such effort. After reviewing some of the logical problems Hempel faced, Goodman finally concludes that "confirmation of a hypothesis by an instance depends rather heavily upon features of the hypothesis other than its syntactical form". ([12], p. 73) The feature at issue concerns the lawlikeness of hypotheses.

Only a statement that is *lawlike* – regardless of its truth or falsity or its scientific importance – is capable of receiving confirmation from an instance of it; accidental statements are not. Plainly, then, we must look for a way of distinguishing lawlike from accidental statements. ([12], p. 74)

If we can distinguish lawlike from accidental generalizations, then we can identify those hypotheses which can be confirmed by instances of them – given (a) that "inductive logic as Hempel conceives it is concerned primarily with ... [the] ... relation of confirmation between statements", ([12], p. 69) (b) "that a hypothesis is genuinely confirmed only by those of its consequences that are instances of it in the strict sense of being derivable from it by instantiation", ([12], p. 69) and (c) that only lawlike hypotheses are, strictly speaking, confirmable.

His program for determining the sense of 'lawlike' is part and parcel of his general view of the status of such legitimizing aspects of other reasoning processes embedded in our language. Consider first his account of the justification for the *rules* of deduction.

The validity of a deduction depends not upon conformity to any purely arbitrary rules we may contrive, but upon conformity to valid rules .... But how is the validity of rules to be determined? ... Principles of deductive inference are justified by their conformity with accepted deductive practice. Their validity depends upon accordance with the particular deductive inferences *we actually make and sanction*. If a rule yields unacceptable inferences, we drop it as invalid. Justification of general rules thus derives from judgments rejecting or accepting particular deductive inferences. ([12], pp. 66–67, my italics)

This is Goodman's famous 'virtuous' circle. We adopt the rules we need to legitimize the inferences we want to make. Whenever we assert that *A* is a rule we do so because it meets our expectations with respect to what rules are supposed to do. In attempting to define 'lawlike' Goodman develops this same line of reasoning. A statement is a law if it does what lawlike statements are supposed to do.

In identifying lawlike hypotheses, however, we do not produce a definition of valid induction. Lawlike statements are universal claims, where Goodman's sense of 'universal' is captured by appeal to complete generality. For instance, in characterizing the problem of distinguishing lawlike from accidental generalizations he says:

The most popular way of attacking the problem takes its cue from the fact that accidental hypotheses seem typically to involve some spatial or temporal restriction, or reference to some particular individual. They seem to concern the people in some particular room, or the objects on some particular person's desk; while lawlike hypotheses characteristically concern all ravens or all pieces of copper whatsoever. Complete generality is thus very often supposed to be a sufficient condition of lawlikeness; but to define this complete generality is by no means easy. ([12], pp. 77–78)

In the ensuing discussion Goodman considers some of the difficulties in defining complete generality, arrives at no answer, but at the same time does not deny that generality is a sufficient condition of lawlikeness. Now, even assuming both that it is a sufficient condition and that lawlike hypotheses are completely general, while we may solve the problem of confirmation by identifying lawlike hypotheses, nevertheless this would hardly appear to have anything to do with induction. Typical inductive inferences are (1) the move from a set of singular statements to another singular statement, or (2) the move from a set of singular statements to a general statement. However, the sort of case with which Goodman deals concerns the determination of a valid prediction, a move from a general statement to a singular statement, hence his concern with laws and generality. This is characteristically a deductive move, albeit sometimes incomplete.

Goodman is not, however, attempting to turn induction into deduction. The relation between the (new) problem of induction and the confirmation of hypotheses lies in ascertaining the legitimacy of the move from a statement known or believed to be true to the claim that some other state of affairs is or will be the case. As Goodman sees it this essentially constitutes the problem of projection and it affects a general area of human reasoning.

Now the problem of making the projection from manifest to non-manifest cases is obviously not very different from the problem of going from known to unknown or from past to future cases. The problem of dispositions looks suspiciously like one of the philosopher's oldest friends and enemies: the problem of induction. Indeed, the two are but different aspects of the general problem of proceeding from a given set of

cases to a wider set. The critical questions throughout are the same: when, how, why
is such a transition or expansion legitimate? ([12], p. 57)

That is, the general activity of making inferences, the basis for which are
past knowledge, past experience or the immediately perceivable, is to be
called 'projection', and the problem is to define 'valid projection'.

This problem, however, cannot be the justification of the *use* of ampliative inference when reasoning about the world. We must recognize inductions
as instances of a legitimate mode of inference. Otherwise the very attempt
to sort valid from invalid projectures would be meaningless. Though *deduction* as a mode of inference is legitimate, not all putative deductions are valid.
The same is true for induction. And, if we push the analogy with deduction,
since valid deductions, or more precisely, deductions strictly taken, conform
to the rules of deductive logic, a definition of valid projection would be in
terms of the rules for projection. And to formulate these rules we must take
more into account than only considerations governing confirmation.

. . . while confirmation is indeed a relation between evidence and hypotheses, this does
not mean that our definition of this relation must refer to nothing other than such
evidence and hypotheses. The fact is that whenever we set about determining the validity
of a given projection from a given base, we have and use a good deal of other relevant
knowledge. ([12], p. 87)

By including semantical and pragmatic elements in our definition, Goodman
argues for a 'reorientation' in our approach to defining 'valid projection'.

I think we should recognize, therefore, that our task is to define the relation of confirmation or valid projection between evidence and hypothesis in terms of anything that
does not beg the question, that meets our other demands for acceptable terms of explanation, and that may reasonably be supposed to be at hand when a question of
inductive validity arises. This will include, among other things, some knowledge of
past predictions and their successes and failures. . . . thus what I am suggesting is less a
reformulation of our problem than a reorientation: that we regard ourselves as coming
to the problem not empty-headed but with some stock of knowledge, or of accepted
statements, that may fairly be used in reaching a solution. ([12], pp. 88–89)

Approach the problem armed with anything that does not beg the issue.
Among these items is our stock of knowledge concerning past regularities.
Other items must include the goals for which we use our knowledge, for
knowledge is always of something with some end in mind.

Goodman's theory of projection begins with actual projections.

Hume thought of the mind as being set in motion making predictions by, and in accordance with, regularities in what it observed. This left him with the problems of differentiating between the regularities that do and those that do not thus set the mind in motion. We, on the contrary, regard the mind as in motion from the start, striking out with spontaneous predictions in dozens of directions, and gradually rectifying and channelling its predictive processes. We ask not how predictions come to be made, but how – granting they are made – they come to be sorted as valid and invalid. ([12], pp. 89–90)

He first characterizes projecting hypotheses and elicits from this a means of determining which ones can be used to make valid projections. These are the projected hypotheses with 'entrenched' predicates. In the case of conflicting hypotheses: favor initially the one using the more entrenched predicates.

An hypothesis actually projected is one actually used. The reasons for its use are not in question – they concern the problem of *justifying* the use of a given hypothesis, which could be viewed as the *'old'* problem of induction.

Actual projection involves the overt, explicit formulation and adoption of the hypothesis – the actual prediction of the outcome of the examination of further cases. That the hypotheses could – or even could legitimately – have been projected at this time is at this stage beside the point. ([12], p. 90)

But to say of a hypothesis that it is actually projected requires, if we are concerned with projecting and not deducing information, that we distinguish between those hypotheses which we use in projecting and those we use for other purposes, e.g., explanation. To this end, Goodman introduces the following notions to differentiate between hypotheses, instances of hypotheses, and other classes. First, he distinguishes between the *positive* and *negative* and *undetermined* instances of a projected hypothesis. He then distinguishes the evidence class from the projective class of an hypothesis.

If the hypothesis is to the effect that all so-and-sos are such-and-such, then so-and-sos named in its positive cases constitute the *evidence class* for the hypothesis at that time, while the so-and-sos not named in either its positive or its negative cases constitute its *projective class*. ([12], p. 92)

Finally, the hypothesis is *supported, violated,* or *exhausted.* A supported hypothesis has positive instances and a violated one has negative instances on its record. Now

... if a hypothesis has both positive and negative cases at a given time, it is then both supported and violated; while if it has no cases determined as yet, it is neither. A hypothesis without any remaining undetermined cases is said to be *exhausted.* ([12], p. 92)

With this preliminary account in hand, Goodman proceeds to the problem of distinguishing lawlike or legitimately projectible from non-projectible hypotheses. He considers a case in which two competing hypotheses, neither of which is violated or exhausted, provide grounds for a projection. How do we decide which of the two to sanction?

The method, briefly stated, requires that we project the hypothesis whose predicates are better entrenched. An entrenched predicate has been habitually projected. One predicate is better entrenched than another predicate for at least the following reasons: (a) it has been projected for a long time, i.e., its use involves a stronger linguistic habit; (b) the consequences of not projecting it over another predicate would have too drastic a set of results for other linguistic habits.

One result of taking both of these conditions seriously is that entrenchment does not automatically accrue to a predicate simply because of habitual use. Goodman distinguishes between entrenchment and similarity. His guard against viewing entrenchment in terms of similarity, and, at the same time his protection against restricting science to the conservatism of habit, is twofold. First, while starting from actually entrenched predicates (for illustrative purposes as well as holding to his procedure of not justifying what we do, but characterizing the differences between legitimate and illegitimate projections), he does not limit the concept of entrenchment to only actually projected predicates. He expands his account to include predicates coextensive with entrenched ones.

The entrenchment of a predicate results from the actual projection not merely of that predicate alone but also of all predicates coextensive with it. In a sense, not the word itself but the class it selects is what becomes entrenched, and to speak of the entrenchment of a predicate is to speak elliptically of the entrenchment of the extension of that predicate. ([12], pp. 95–96)

Thus an unfamiliar predicate can be said to be entrenched if it can be shown to be coextensive with a familiar entrenched predicate. The pragmatic observation that we must be allowed to consider unfamiliar predicates if knowledge is to progress constitutes the second point.

... any wholesale elimination of unfamiliar predicates would result in an intolerable stultification of language. New and useful predicates like 'conducts electricity' and 'is radioactive' are always being introduced and must not be excluded simply because of their novelty. ([12], p. 97)

At this point, however, I want to raise a rather important question. New and useful predicates are constantly being introduced, but useful for what? Consider another claim by Goodman:

In framing further rules, we must continue to be on guard against throwing out all that is new along with all that is bad. Entrenched capital, in protecting itself, must yet allow full scope for free enterprize. ([12], p. 97)

Why must capital allow for free enterprize? Obviously, in economic terms, to permit economic development, i.e., progress. But one of the objections being raised today against 'economic progress' and its handmaiden 'increased productivity' is that there appears to be no reason behind the continued rush to expand economic growth now that the original major economic objectives of enough goods and sufficient material items for comfort have been met.

What then does science seek? If the theory of projection must be on guard against excluding what is new, Goodman must be at least nominally committed to the idea of scientific progress. But he limits his concern to obtaining what is useful He leaves room for 'progress' only because the class of all possible problems we have to meet has not been exhausted. We must leave room for new ideas because new problems may require new solutions based on as yet unthought of concepts. This would be compatible with Feyerabend's extremely pragmatic method, theoretical pluralism. Under these circumstances we would be well-advised to encourage the development of as many different theories as possible. This way we could be stockpiling solutions against future problems.

All of this is very sketchy. The point at issue, however, does not require a great deal of embellishment. Are we to characterize science as progressing toward some specific end or do we measure scientific progress in terms of an increased ability to solve problems? This too, however, is an end. The difference is between science pursuing an end such as truth, or an end less absolute in an important way. If truth is our objective, then, following Sellars, we have the beginning of a justification for change. If we can in some way determine that our theory does not give us true explanations, then we have good reason to change the theory and to continue changing it until we get only true explanations. But, to be momentarily misleading, Sellars' notion of truth here is ontological. A *theory* is true if it accurately pictures the universe. Hence, when we develop that theory which gives us the complete ontological structure of the universe we will have achieved our end; we will have the truth.

But, if, like Goodman, we assume that absolute truth is not forthcoming, nor even possible, then our reason for abandoning a theory or seeking changes is not as obvious. Goodman seeks to insure the possibility of change so that we may discover something useful. It can be assumed then that once we have reached a plateau of some kind we will no longer have reason to seek new ideas. If no theory giving us an absolute view of the universe can be had, then we can stop looking when we have a theory which allows us to solve all our problems. And if there are no unsolved problems, then apparently we have no reason to inquire further into things. This approach can be characterized as pragmatic. If solving problems constitutes an objective of science then, the admonition goes, do only what you must to get us to the point where we can solve those problems. But, once we have reached that point, there is no reason for proceeding toward anything like an ideal of truth (unless, of course, truth is an epistemic goal and epistemic goals also count).[3]

If a choice exists here, if we can choose one or the other as the ideal, then the following questions need to be considered.

(a) Are there any arguments which can be offered in favor of a positive response to one and a negative response to the other?

(b) Are the two alternatives incompatible?

In response to the second question, we can say that the two alternative characterizations are not in principle incompatible, though traditionally they have been so considered. Sellarsian scientific realism constitutes a synthesis of both views. Sellars argues that progress in science is to be viewed in terms of the development of a completed science. The goal, however, is not truth, but explanatory coherence. A complete ontology constitutes one result of this approach. The fruitfulness of any particular ampliative move in contributing toward a more complete and comprehensive analysis of the world becomes its justification. I characterize this view as epistemologically pragmatic and ontologically realistic.

The more traditional approach assumes the two views are concerned with different issues. If science culminates in some end such as a complete account of what there is, then the concern of those analyzing science in this way is obviously ontological. On the other hand, if science remains only a tool in problem solving, then the concern is epistemological. But, one of the crucial problems in theory construction, i.e., the introduction of theoretical terms, seems completely answerable only if we commit ourselves to an ontological position and use it to guide the epistemological investigation.

This issue, however, takes us back to the first question posed above. Are there any arguments in favor of one view over the other? The claim that only on, for instance, a realist ontology can the introduction of theoretical terms be fully justified constitutes a good reason for arguing that science must be viewed as striving toward a goal, and in particular, an ontological rather than an epistemological goal. Otherwise the interest of science in systematization is hard, if not impossible, to explain.

Goodman and Quine, however, appear to be headed in a different direction. They seem convinced that the solution to the epistemological issue of entrenchment will permit the solution of certain ontological problems, thereby arguing for the problem oriented view of progress. Consider their approach to the notion of natural kinds.

Goodman expects a solution to the question of kinds to result from the development of a theory of projection. His solution to the problem of confirmation, or the validity of inductive inferences, rests on the concept of entrenchment. Approaching the problem from this angle requires taking the linguistic dimension of the issue seriously. And Goodman is quite explicit about this point. Projectible hypotheses are identified by their use of entrenched predicates and "entrenchment derives from the use of language". ([12], p. 96) But to state only that does not clarify the issue, for surely the sense of 'derives' is metaphorical here. Goodman elaborates:

If I am at all correct, then, the roots of inductive validity are to be found in our use of language. A valid prediction is, admittedly, one that is in agreement with past regularities in what has been observed; but the difficulty has always been to say what constitutes such agreement. The suggestion I have been developing here is that such agreement with regularities in what has been observed is a function of our linguistic practices. Thus the line between valid and invalid predications (or inductions or projections) is drawn upon the basis of how the world is and has been described and anticipated in words. ([12], p. 117)

Thus entrenchment remains a feature of the language we use. And once having recognized that valid projections are determined by the content of our language, we can reasonably ask the next question: What is the relation between our language, the content of which is evidenced in entrenched predicates, and the world? While Goodman does not ask this specific question in *Fact, Fiction, and Forecast*, he does direct some attention to the issue there.[4] More precisely, he considers using the theory of projection to distinguish 'genuine' from 'artificial' kinds.

Our treatment of projectibility holds some promise in other directions. It may give us a way of distinguishing 'genuine' from merely 'artificial' kinds, or more genuine from less genuine kinds, and thus things are or are not of the same kind, or are more akin than certain other things. ([12], p. 119)

There are two important things to note here: the meaning of the term 'kind' and Goodman's inadequate proposed epistemological solution to an ontological problem. The 'kinds' Goodman speaks of here are readily recognizable as universals. This is clear from his description of the role they play. They are classificatory predicates which permit the regulative description of items in terms of their similarity or dissimilarity with respect to the characteristic which the predicate names. Thus, for example, if the predicate in question is 'red', two objects can be described *vis-à-vis* their being or not being red; they can be grouped in the class of all red objects if both are red, or in the class of all non-red objects, or neither are red, etc.

Now, given the role of such predicates, the status of the 'item' which they name, i.e., redness, remains to be determined. Is there something in the world which is *red simpliciter*?

Goodman's desire to separate off artificial from genuine kinds raises the initial question as to what he could mean by 'genuine'. There are two possible answers. Either they are genuine because they are well entrenched predicates of our language and have long established classificatory roles, as opposed to transitory *ad hoc* classificatory devices, such as 'railroad conductor', or they are genuine because the item mentioned by the predicate exists in the world and is not merely the product of conventions like languages.

To suggest that Goodman should be read as a realist simply because he makes an obscure reference to 'genuine' kinds should raise objections. Goodman's nominalism is well known, and his sense of realism can only, if at all, be limited to the insistence on individuals. I am not, however, claiming Goodman is a realist, but rather that he can be read here as manifesting realist tendencies, which impression he hastens to explain. He weakens 'genuine' by speaking of 'less artificial' and 'more genuine' so as to comfort those readers who might be shocked by his apparent presuppositions that there are, in fact, kinds and moreover, that we can come to know them.

But without realism Goodman's hope that the theory of projection might solve the problem of kinds and along with it "questions concerning the simplicity of ideas, laws, and theories" rests on a circular argument. ([12], p. 119) Consider first Goodman's suggestion as to how the theory of projection can solve the problem:

For surely the entrenchment of classes is some measure of their genuineness as kinds; roughly speaking, two things are the more akin according as there is a more specific and better entrenched predicate that applies to both. ([12], p. 119)

The more entrenched the predicate the closer it comes to being a genuine kind. Entrenchment constitutes the criterion for genuineness.

But, if predicates become entrenched only through habit, if there is no deliberate process of selection, then no *justification* for claiming that a predicate which has managed to survive is a genuine kind can be offered. To merely say that if it survives it must be a genuine kind is not enough. The possibility of a predicate surviving as a result of an epistemological accident remains very much alive. And in this sense the old problem of induction raises its ugly, but for Goodman, senseless head.

Goodman puts his point about the relationship between projectible predicates and genuine kinds in the following way.

I submit that the judgment of projectibility has derived from the habitual projection, rather than the habitual projection from the judgment of projectibility. The reason why only the right predicates happen so luckily to have become well-entrenched is that the well-entrenched predicates have thereby become the right ones. ([12], p. 98)

Now clearly if habitual projection is the manner of genesis of entrenched predicates, then it hardly follows that entrenchment will provide the clue to the genuineness of kinds. At best we have an incomplete argument which needs an additional premise such as: the survival of a predicate must be an indication of its corresponding to some entity in the world. At worst, the argument is flagrantly and perhaps viciously circular. Some statements about the world give us knowledge because they use entrenched and projectible predicates and we know the predicates tell us about the world because they are entrenched.

This is not to claim that from Goodman's vantage point the account is incomplete. Rather, I am suggesting that talk about the 'genuineness' of kinds, with some relation between our linguistic framework and the world is peculiar to say the least. It is one thing to say that the system, by virtue of its history, allows us to use predicates in specified ways, and quite another to say that because the system allows us to use predicates this way they represent 'genuine' kinds or types. The 'genuineness' of kinds is a function of the 'real' order of the world, not the language.

Now this kind of discussion is awkward because the charge of incompleteness against Goodman's account can simply be dismissed as missing the point. Goodman can be reasonably read as arguing for nothing more controversial than a method for developing a coherent system, wherein the predicates of the system can be used in particular ways because they have been so used and we restricted their use by the conventions of the language, which conventions have been worked out over a period of time.

But the whole point of arguing that Goodman shows realist tendencies and

that his argument is either incomplete or circular is to build up the idea
that there is more of importance here than meets the eye. The weakest
formulation of the claim maintains that talk of genuineness is inappropriate.
Natural kinds are a matter of convention. They are genuine on 'natural'
only to the extent that they represent the seasoned development of our
system.

But I am really concerned with the stronger claim: the *assumption* that
there is some link up between our conventional language and the way the
world really is stands as a logical *prerequisite* for the continued development
and manipulation of our language, *if* that development is to be justifiable.
Goodman's talk about genuine kinds is important because it provides a good
example of someone not committed to realism who nevertheless still feels the
importance of something absolute or stable if only as a goal, i.e., something
to shoot at.

Goodman backs off, qualifies his position, reestablishes his credentials
as a pragmatic nominalist and so, if we are not only sympathetic but also
go into other works of his, completely clears himself of any charge of realism.
That, by itself, does not mean he is out of the woods. For the objection to
this kind of isolationism still holds. One of the primary considerations in
developing a system supposedly about the world involves the assumption of
a tie between the system and its object. Arguing the possibility of such a
tie misses the point. That it is our *goal* remains crucial, for without it all
talk about 'laws of nature' and theories of the 'universe' becomes perverse.
In other words, in a system designed to accommodate change, part of the
justification for such change must entail the idea that this will bring us closer
to our objective: a framework in which we can provide true descriptions and
explanations of the phenomena we observe.

Goodman does not attempt to resolve the circle by any claim of virtuosity
here. In fact, he does not consider the possibility that the solution to the
problem of kinds *via* the theory of entrenchment might be too circular even
for him. But if he insists on the tenability of this position, then he must
either widen the circle or explain it away.

Quine picks up the first alternative noted above and supplies a missing
premise by appeal first to genetics. He then claims that while entrenchment
picks out kinds, entrenched predicates are related to *natural* kinds more by
virtue of our biological makeup than by corresponding to necessary features
of the world. He views the task of science as the elimination of kinds in this
sense.

Quine's attempt to account for natural kinds in terms of biology does not, however, produce an adequate solution to the problem of their epistemological status. It does not tell us what kinds or universals are.

4. QUINE, INDUCTION, AND NATURAL KINDS

Quine begins by relating the problem of confirmation to the characterization of projectible predicates in the Goodman manner. He argues that at bottom the projectibility of predicates rests on or requires the prior concept of similarity. To project a predicate on the basis of past experience requires grouping those past experiences as similar and assuming that future instances will be similar to the past instances. Then, in the context of discussing the grue-predicates, he introduces the role of kinds.

A projectible predicate is one that is true of all and only the things of a kind. What makes Goodman's example [i.e. grue] a puzzle, however, is the dubious scientific standing of a general notion of similarity, or of kind.

The dubiousness of this notion is itself a remarkable fact. For surely there is nothing more basic to thought and language than our sense of similarity; our sorting of things into kinds. The usual general term, whether a common noun or a verb or an adjective, owes its generality to some resemblance among the things referred to. ([32], p. 116)

The problem, i.e., the dubiousness, arises because even though the two concepts of similarity and kind are fundamental to our thought processes, we cannot define them independently. Quine first argues that similarity is not completely definable in terms of kind or vice versa.

Still the two notions are in an important sense correlative. They vary together. If we reassess something *a* as less similar to *b* than to *c*, where it had counted as more similar to *b* than to *c*, surely we will correspondingly permute *a, b,* and *c* in respect of their assignment to kinds; and conversely. ([32], p. 121)

He next distinguishes between properties and kinds, where kinds are characterized extensionally and properties intensionally, by initially attempting to identify kinds as sets.

The contrast between properties and sets which I suggested just now must not be confused with the more basic and familiar contrast between properties, as intensional, and sets as extensional. Properties are intensional in that they may be counted as distinct properties even though wholly coinciding in respect of the things that have them. There is no call to reckon kinds as intensional. Kinds can be seen as sets, determined by their members. It is just that not all sets are kinds. ([32], p. 118)

This very difficulty, i.e., that not all sets are kinds, leads Quine to the conclusion that similarity cannot be defined in terms of kinds. And, of course, the problem should have been evident from the start if kinds are defined as sets. We identify an object as belonging to a set by attributing to it a property which captures the character of the set. And Quine is quite right when he notes the difficulty in defining 'property'. His problem stems, however, from the methodological presupposition that definitions of key concepts must be extensional.

Let me suggest a different approach. The notions of kind and similarity are indeed fundamental. The way to capture the fundamental role that these concepts play in our thought process, however, involves characterizing them in terms of constitutive rules which introduce them into our own conceptual framework. They are part of the basis for thinking about objects. Support for this claim can be found by considering, for example, the subject-predicate character of our language. Reflecting on our discussion of Sellars, note that formation rules of a language are the basis for correct grammatical criticism, where criticism is construed both positively and negatively. But the 'rules' behind the formation rules tell us something different: how we ought to think about the world. In particular they tell us that when we describe the world, we say of an *object* that it *has* some *property* or other. Formation rules therefore require some criterion by which we can determine whether or not they are doing their job. The criterion is a high degree of explanatory coherence.

We must assume the world has some specific character before we begin to describe it and construct rules concerning the proper and improper form for descriptions. This assumption, captured by the subject-predicate form of our language, is constitutive with regard to our thought about the world. That there are objects and properties constitutes the assumption in question. Without the assumption of the *existence* of properties talk about objects would be impossible. Whether *or not* properties actually exist in the world, predicates naming the properties play a fundamental role in our thinking about the world, so fundamental in fact that it would be impossible to describe the world without their use. Thus it is a mistake to define properties extensionally because predicates naming those properties are a constitutive part of our language. To understand what a property is requires understanding the role predicates play in language. The assumption that there *are* properties remains embedded in our *language* and, hence, our way of thinking.

But what ontological status do properties have? There are at least three possible answers: Goodman's, Sellars', and Quine's. Goodman argues that in effect they have no status; begin the analysis of predicates by considering which ones have managed to survive. Whether or not they refer to some entity in the world or whether they mark some feature of the relationship between Man and his world are interesting, but unimportant, questions. Sellars feels that the subject-predicate form of language provides an important clue to the way things really are and moreover, the development of science involves the construction of increasingly sophisticated predicates as a means to isolate whatever we might eventually come to finally accept as a property. Quine starts off by agreeing with Sellars that the subject-predicate form of language does contain a clue to nature of properties. He analyzes this clue by pursuing two lines of thought. The first is genetic. The second develops the suggestion that while man's genetic character constitutes the source of the subject-predicate language, science is the hallmark of man's ability to transcend his animal nature to overcome this heritage and eliminate predicates, i.e., kinds.

Quine also adopts the Sellarsian presupposition that there is an isomorphism between predicates in languages and properties in the world. Sellars believes that the complete description of the world which a completed science will offer requires, however, that our present predicates be replaced with predicates that provide a better 'picture' of the world while displaying their historical ties to their ancestors. While he believes in the essential correctness of such an isomorphism, he also feels that our present language does not demonstrate it in a way which gives us the correct view of the universe. To arrive at the correct view we must go from what we *can* see to a new and different picture; we must construct new theories and introduce new entities to provide a more complete description than we now have. But changing our picture in this way requires that the process of changing pictures be legitimate, i.e., that projection or induction, be accepted as a legitimate mode of enquiry.

Quine takes a slightly different tact. He agrees about the isomorphism between the predicates in our language and properties in the world. *But*, he qualifies this claim in the following way. The isomorphism obtains between predicates and properties in the world, where the world as experienced is a conjunction of our biological limitations. But the world we see is not necessarily the 'real' world. For example, we cannot perceive ultraviolet or infra-red colors, but this does not mean they do not exist. If we are to analyze color as a product of the interaction of an observer, lightwaves and objects, then a corollary of this theory is that some colors are unobservable. Something follows the same pattern as the one which causes us to perceive red but at the same time it does not cause us to perceive any color. The explanation for this phenomenon is that our eyes have a limited receptivity and this particular interaction of object, wave length and observer such that the observer cannot be an accurate instrument for registering the interaction. The color is there, even though we don't see it; a rather paradoxical claim.

Such paradoxes as these are behind Quine's claim that as natural kinds, colors don't exist. They are biologically-derived-sorting-devices which do not, however, tell us what the world is really like. Support can be found in the product of scientific research. Science tells us there must be colors that we can't see. Color remains a 'natural' kind only as a product of our animal nature. It plays a useful role in the area of survival, but becomes a hindrance when we are concerned with issues beyond mere survival.

Quine recognizes two distinct characteristics of, shall we say, human nature: (1) the set of natural limits imposed on us by our biological makeup (he uses variations on the phrase 'innate quality space' to refer to this feature of our predicament); (2) our natural ability to overcome our innate quality spacing, our innate sense of similarity and its concurrent formulations of predicates naming kinds in our language, i.e., our ability to reason ampliatively. The ability to reason ampliatively constitutes a genetic trait of man.

After characterizing similarity as a biological phenomenon, rather than a linguistic concept which admits of logical analysis, Quine throws induction into the same boat.

Induction itself is essentially only more of the same: animal expectation or habit formation. And the ostensive learning of words is an implicit case of induction. Implicitly the learner of 'yellow' is working inductively toward a general law of English verbal behavior, though a law that he will never try to state; he is working up to where he can in general judge when an English speaker would assent to 'yellow' and when not. ([32], p. 125)

Aside from the overwhelming *similarity* between Quine's views here on the role of induction and Sellars' notion of material moves and rules of criticism, this passage is also important because of the *contrast* it presents to Sellars' position. Sellars views induction as a necessary prerequisite for the inclusion of new material moves and considers it in terms of the logical role it plays. Quine ignores the Kantian epistemological sophistication and says straightforwardly that that is the way we are built: we make inductions.

Now this approach is next to useless. Quine makes an empirical claim. Sellars, on the other hand, offers an analysis of the logical requirement for developing a complete description of the world. As an empirical observation, Quine's claim can *always* be substantiated by arguing that the absence of the allegedly innate characteristic constitutes either an abnormality or further evidence for the original claim itself. This latter possibility occurs, for example, when someone fails to see the same colors we do and we call him color blind. Whatever else, this view certainly sheds no epistemological light.

Now, assuming that Quine can make plausible his neo-Goodmanian analysis of induction, let us consider how he plans to use induction to serve the paradoxical situation noted earlier. Man has two natural qualities, innate color spacing and an innate ability for induction.

Living as he does by bread and basic science both, man is torn. Things about his innate similarity sense that are helpful in the one sphere can be a hindrance in the other. Credit is due man's inveterate ingenuity, or human sapience, for having worked around the blinding dazzle of color vision and found the more significant regularities elsewhere. Evidently natural selection has dealt with the conflict by endowing man doubly: with both a color-slanted quality space and the ingenuity to raise above it. ([32], p. 128)

The conflict Quine has in mind lies between what our senses tell us and what science tells us. Our ingenuity to rise above biological limitations is our way out of the conflict. This involves accepting the results of science and rejecting our primitive observational source of knowledge. Science is the product of this trait which has been cultivated by natural selection.

The genetic 'explanation' offered by Quine constitutes, however, only one aspect of his account of induction. The second, and more important, includes the result Quine envisages as the proper end to the use of our natural inductive talents: the eventual replacement of natural kinds by less universal concepts. We can thereby eliminate the apparent paradoxes that occur when, for example, science tells that there are colors we cannot see.

In principle Quine is quite close to Sellars. They both consider the present conceptual framework inadequate. They both seek to have it replaced. But Quine's view differs in a crucial fashion. It envisages the elimination of the concepts of kind and similarity as general constitutive aspects of our conceptual scheme. They are to be replaced by particular concepts which play the same role but which are peculiar to individual sciences.

Different similarity measures, or relative similarity notions, best suit different branches of science; for there are wasteful complications in providing for finer gradations of relative similarity than matter for the phenomena with which the particular science is concerned. Perhaps the branches of science could be revealingly classified by looking to the relative similarity notion that is appropriate to each. ([32], p. 137)

Both Sellars and Quine agree that science will provide the final answers.

But, for Quine the final science will be fragmented and composed of individual sciences.

Despite some resemblances between Quine's and Sellars' views there are serious difficulties with Quine's account which do not appear in Sellars'. Consider Quine's panacea. By eliminating the general concept of similarity and kind and replacing it with functional definitions appropriate to the particular science in question we not only eliminate the problem of defining the concepts in question, but we permit science to get on with the business of solving its problems. However, as we saw earlier, Quine firmly believes that the concepts of kind and similarity are fundamental. How fundamental he does not appear to fully appreciate.

Without the general concept of kind it would be impossible for a particular science to isolate or identify a problem, or attempt solutions. The termination of their role in the thought process results in the termination of the thought process itself. These concepts are constitutive of thought insofar as they permit the identification of discrete episodes. Without discrete episodes the ability to organize data, a necessary prerequisite to theory guilding and problem solving, would be superfluous since it would have nothing to organize. Two interrelated functions are involved here: the sorting of the haze of sensory exposure into discrete episodes, and what for lack of better terminology I call the *character* of the sorting. Experience involves more than just organizing data on a time-space coordinate system in order to qualify as something more than mere sensation. Paralleling Goodman's 'regularities are where you find them', objects are where you find them. We need a functioning 'kind-discriminator' and a 'similarity-secretary' in order to find objects. The kind-discriminator organizes the sensory information into one sort of thing rather than another and the similarity-secretary notes it as belonging to group A as opposed to group B. Only when these jobs have been done can we use the information and not until then.

Quine's mistake lies in generalizing from the acceptable practice found in theory building, of replacing 'common sense' by theoretical descriptions, to the idea that we can replace anything. While we may sometimes replace one kind by another, that does not permit the further move that the notion of kind itself may be replaced. For if it did, any identity of field as well as categorization of data would be impossible.

In trying to come to grips with the motivation for Quine's move here it seems plausible to suggest that he is worried about the stymieing effect of reductionism in science. As he points out, the development of individual sciences without regard to the problem of other fields has led to several successes, where he defines success in terms of successful problem solving. We might consider, therefore, the possibility that the final science may not be a single deductively coherent system.

The results of this suggestion, however, can be viewed as a *reductio ad absurdum* of the problem-solving view of science. Progress in science results in eliminating the possibility of knowledge, hence problem-solving.

Quine, aware of the problem tries to counter such an objection by vaguely alluding to what would appear to be a Sellarsian final science. After noting that each branch of science should develop its own similarity notion, thereby eliminating the problem of defining kinds, he continues:

Such a plan is reminiscent of Felix Klein's so-called *Erlanger-programm* in geometry, which involved characterizing the various branches of geometry by what transformations were irrelevant to each. But a branch of science would only qualify for recognition and classification under such a plan when it had matured to the point of clearing up its similarity notion. Such branches of science would qualify further as united, or integrated into our inclusive systematization of nature, only insofar as their several similarity concepts were *compatible*; capable of meshing, that is, and differing only in the fineness of their discriminations. ([32], pp. 137–138)

Thus, sciences are admissible into whatever Quine has in mind provided they are of a *kind*. But determining whether or not the various similarity notions 'mesh' requires a standard, an idea of the kind of similarity. As much as Quine would like to eliminate a general concept of kind and similarity, he cannot. To achieve an organic unity among the various branches of science he needs both 'kind' and 'similarity'. Without 'similarity' he cannot determine if the differences between the discriminatory abilities of individual sciences are only in terms of 'the fineness of their discriminations'. Without 'kind' he can't determine if the thing in question is a science or has a similarity notion. He must either eliminate these concepts and a unified body of knowledge along with them, or eliminate one of the concepts and replace it by a notion that won't work, accomplishing the same result.

Quine fails to solve the problem by arguing for the elimination of the general concept of kinds from our conceptual scheme. His efforts here are fundamentally misdirected. By not recognizing the constitutive nature of 'kind' he undertakes an ultimately self-defeating program. This is not to say that he was wrong in his intuitions. Indeed it may very well be helpful to realize that certain predicates we have used to determine class membership ought no longer to be so used. The problem is how to cash in this insight without falling into the same sort of trap Quine did.

I believe I have at least the outline of a solution here. What I hope to

demonstrate, however, is not a final account of 'kind', but rather what else
is needed to complete the theory of projection. The excursion into Quine's
account was intended to be illustrative of the general point that something
fundamental is missing. To complete the picture let us return to Sellars' rules
of inference.

In effect rules of inference govern the use of a concept as it is employed
in object-language expressions. To say a predicate is entrenched is to say
that its use in some object-language expression is well-governed. This is
to say that the rules concerning its use are well-developed. In other words
to say a predicate is projectible is to say that the rules governing its use have
been accepted as constituent members of the meta-language.

The purpose of explicating 'entrenchment' and 'projectible' in terms of
meta-linguistic rules of inference, is to help clarify the semantics and prag-
matics of evidence. The considerations of relevance and specification of truth
conditions for the proper application of a predicate are surely matters to be
settled at the meta-level. But the actual employment of the predicate in a
confirming or diconfirming role is an object-level activity. One reason Quine's
efforts to follow through on Goodman's suggestions about 'kind' did not
succeed was his failure to recognize the fact that 'kind' is not object-level
predicable. It cannot operate at the object level in the way a projectible
predicate can. In other words, the meaning of the predicates is a function of
their use. How they are to be used, as opposed to how they are used, is
determined by means of which rules of inference govern them. The determination of the proper application of a predicate is a matter of finding out
if the use of the predicate has been properly inferred from the rules. This
is a matter of deduction. In this case Quine incorrectly inferred that 'kind'
was a projectible predicate.

One way of summing up is to observe that induction has nothing to do
with evidence. It is constitutive of inventiveness and creativity. Or to put
the point in a linguistic context congenial to Quine, induction is constitutive
of a language capable of evolving concepts. Thus, returning to Quine's example, if it were clear what role the sorting predicates were to perform for a
given science, then the independent development of those sciences could be
encouraged without the gross action of eliminating kinds in general. That
is, it is a meta-level decision as to how to use a given predicate and the distinction between different uses is quite possible once you know where to make
it.

To turn to the second qualification, projectibility should be tied to the
concept of evidence. An observation report counts as a piece of evidence if

and only if the predicates contained in the report are entrenched. This entails that negative reports render the rules less well-developed. The justification of a generalization proceeds not by citing particular data, but by citing reasons, i.e., the rules for generating particular observations. If a rule is shaky, it cannot be unqualifiedly used as a reason.

5. CONCLUSION

I have tried to show that we can distinguish between generating new ideas and justifying their use by first reconsidering our analysis of induction. If induction is a mode of reasoning, then there is no need to justify it. The problem of justifying particular inductive inferences, however, remains. Using Goodman's theory of projection, I have sketched, all too briefly, an account of how this can be done without encountering the problems associated with induction. Essentially the point is that justification proceeds by citing reasons for conclusions. This point is amplified in the next two chapters. Observation statements are not reasons for anything. They are assertions of fact or reports. In either case they can be used as data provided their predicates are well-entrenched. Entrenchment in this case is a two-fold notion. It involves not only projectible predicates, but also agreement on the rule governing the predicate. Thus a predicate can have functioned most respectably but still be abandoned if it is agreed that we no longer want to do the sort of thing that predicate has been used for. This approach has an interesting result. If characterized in this fashion, the conceptual scheme in which scientific investigations are carried out admits of the possibility of changing views on the nature of its own goals. In which case it admits of room for disagreement over goals. Such an account is neutral, therefore, with respect to realist and nominalist, despite the fact that I have played out some realist sympathies.

CHAPTER III

INDUCTION AND JUSTIFICATION

1. INTRODUCTION

Sellars' theory of justification rests on a theory of probability wherein a reorientation is urged on us concerning the meaning of 'probable'. To say a statement is probable, means, according to Sellars, that there is good reason for accepting the statement as true. The giving of good reasons involves practical reason and a general goal. In the context of science practical reasoning occurs in the meta-language of a theory. The goal is an explanatorily coherent theory.

The acceptance of empirical generalizations (universal or statistical) formulated on the basis of insufficient data constitutes a problem for any theory of scientific reasoning. Sellars uses the straight rule of inductive acceptance. He attempts to deflect traditional objections by embedding its use in a vastly more complicated system than previously used. In this context the distinction between empirical generalizations and laws of nature becomes crucial. Using the object-language/meta-language distinction, Sellars amplifies the difference between rules and laws (laws of nature are rules of inference). The straight 'rule' of inductive acceptance functions only at the object-language level. Generalizations formed in accordance with it are justified to the extent that they agree with the facts. Use of the generalizations for the purpose of explanation, on the other hand, involves rules of inference. The justification of rules of inference differs substantially from the justification of empirical statements. Rules of inference are justified if they contribute to explanatory coherence. Thus, the acceptance of a rule of inference at any given time can only be tentative. Since we have no permanent set of rules (no complete final theory) it cannot be known in advance if any given rule will add to the explanatory power of a theory. Working within a theory, a newly formulated rule of inference already has initial probability, i.e., good reasons for acceptance, because it is, in a sense, already conceptually conditioned. It is rejected when it produces faulty inferences.

Sellars' theory of justification entails, therefore, a bifurcation of the total structure known as a theory as well as the problems of justification. His use of practical reasoning constitutes a unifying device. But to fully appreciate the scope of Sellars' system we need to go beyond what he says and see practical reasoning in an even larger context. And so in what follows I first sketch out the broad structural relations between theories and conceptual frameworks before going on to a more complete analysis of problems of justification produced by Sellars' discussion.

2. RULES, THEORIES, AND CONCEPTUAL FRAMEWORKS

A rule is always a rule for *doing* something. In other words, any sentence which is to be the formulation of a rule must mention a doing or action. It is the performance of this action (in specified circumstances) which is enjoined by the rule, and which carries the flavour of *ought*. ([43], p. 329)

We often call Modus Ponens a rule of inference. Inferring constitutes the action it permits. But Sellars claims it is inappropriate to call an inference schema a rule.

$$P \to Q$$
$$\frac{P}{Q}$$

is an example of Modus Ponens, where:

(MP)   Modus Ponens = df . . . . Given both $P \to Q$ and P then one ought to conclude Q.

Lawlike propositions are, like MP, rules of inference. They are meta-linguistic claims about propositions in a theory T, or more specifically, about the inferential relation between propositions in T. If we divide T into two parts: an interpreted calculus, T_{ci}, and a set of rules, T_r, we can further characterize T_{ci} as the object-level of T, and T_r as the meta-level. Now the conceptual form of a lawlike statement is roughly indicated by the following example:

For all temporal senses t, one ought not to accept both the proposition that there is lightning at t and the proposition that there is not thunder at t plus Δt.

This is, in the first approximation at least, equivalent to

(t)   that there is lightning at t implies that there is thunder at t plus Δt where 't' ranges over the appropriate temporal sense or intensions. Thus lawlike statements are at the meta-linguistic (and meta-conceptual) level, and must be carefully distinguished from qualified statements at the first level of discourse. As indicated, they involve quantification over intensions or senses. Thus the above implication statement must not be confused with the object language statement:

(t)   there is lightning at $t \rightarrow$ there is thunder at t plus Δt where '$\rightarrow$' stands for material implication, and 't' ranges over moments of time. ([48], p. 117)

Lawlike statements are ought statements. But they are also rules of criticism.

... lawlike propositions tell us how we ought to think about the world. They formulate rules of criticism, and if, as such, they tell us what ought or ought not to be the case, the fact that it is what ought or ought not to be the case with respect to *our beliefs about* the world suffices to distinguish them from those rules of criticism which tell us what ought or ought not to be the case in the world. ([48], p. 117)

A further distinction between descriptions and our use of descriptions is also involved here. Descriptions of the world are those statements which tell us what the world looks like. These can be verified or confirmed. Ought-statements are rules of criticism about our use of descriptive statements. They tell us how we ought to *think* if we are to produce an explanation, meet success in our experiments, etc. More generally, ought-statements such as the above example of the conceptual form of a lawlike statement, are rules of inference. They tell us what we ought to do, namely: do not accept a statement, x, and another statement y.

As it stands, however, this account is incomplete. It leaves unmentioned the overall reason why we 'ought' to think this way, a consideration incorporated in Sellars' theory of practical reason. The goal is to produce an explanatory framework. Only in referring to that goal does talk about laws as ought-statements make any sense.

However, referring to such a goal requires enlarging our prior description of a theoretical framework. We now need a three fold distinction between an object-level, a meta-level and a conceptual framework. T_{ci} includes the physical-thing language. The meta-level is the framework T_r in which we locate rules which tell us what to believe in the context of a theory. The conceptual framework *CF* is the context in which we formulate general policies, goals, and rules governing particular areas of inquiry. *CF* must be capable of mentioning liguistic pictures as well as containing rules for manipulating them if we are going to be able to criticize rules. We must criticize the rules or we will not be able to change them in those cases where they give us bad theories.

Lawlike propositions are, therefore, rules of criticism concerning our use of descriptive statements. They guide our reasoning. By allowing that we ought not to accept both x and y, we know that we may *not* infer y if given x. And in this sense lawlike propositions are rules of inference.

In this context we can, among other things, show how Sellars' account of lawlike statements explicates Goodman's concept of entrenchment. On the surface it bears little resemblance to the Goodman style where an entrenched predicate has its use well governed. A concept is well governed when the rules governing its role in the language are sufficiently well developed and tested to permit a number of inferences to be made. The more inferences which can be made with respect to that concept, the more developed it is.

But while Goodman was on the right track, he was at the wrong end. We should not be concerned with discovering rules to identify entrenched predicates in the hope that they will provide us with the criterion for lawlike statements. It is the other way around. Lawlike statements provide us with candidates for entrenched predicates. An entrenched predicate is well integrated into our reasoning about a given domain in such a way as to permit a consistent and coherent use. It is so well integrated in fact that we are less willing to give it up than other not so well entrenched predicates. The full force of this development cannot be felt, however, until we specifically speak to the issue of conceptual change in Chapter IV.

We have here an account of lawlike propositions which attempts to characterize the relation between rules of inference and statements in the object-language. It differs from Sellars' by insisting on an extra conceptual level where general goals and general rules can be formulated. Reconsidering our intitial characterization of T, let us talk instead about a theory as a multi-leveled system with a set of lawlike propositions at the meta-level used to manipulate statements in the object-level organized into a system of material implication. In this context the requirement for the third level, CF, is even more obvious. For only in CF can we think *about* the primitive concepts of T. Constitutive rules such as those licensing induction must also be located there. Using induction to construct a theory depends on the availability of a generic rule of induction located in CF. It constitutes our ground for characterizing empirical science as ampliative. Inductive reasoning concerning matters of fact, i.e., internal theoretical reasoning, is a specific application of the generic principle of induction. A specific use of induction within a theory presupposes such a generic principle. This amounts to a transcendental justification for induction, all that we can provide.

By placing Sellars' distinction between laws of nature and general descriptions within a general framework, I have tried to show that *CF* is a necessary adjunct to the Sellarsian view. As a justification for the legitimacy of making inductive inferences I propose something akin to a transcendental reason: without induction we could not do science. Instead of filling this out to form a deduction as Kant might have preferred, I have simply taken a line from Sellars' program and attempted to use it in a manner consistent with his approach. Given this description of the general framework within which scientific reasoning occurs we can now turn our attention to Sellars' theory of probability.

3. JUSTIFICATION, PROBABILITY, AND ACCEPTANCE

Theoretical justification concerns the giving of reasons. In Sellars' characterization of a theory, *T*, its formal structure lies within a framework of practical reasoning which provides the mechanism for using the theory.

What are the reasons for accepting the conclusion of an argument formulated within *T*? Why and in what circumstances should we accept as true statements derivable from some given set of acceptable premises? No problem exists for deductive derivations. Given a sound derivation, our acceptance of the conclusion of the arguments is justified by our prior acceptance of the rules governing deductive arguments. Once we spell out the form of the type of practical reasoning we shall see that the use of the rules of deductive logic can be classified under the rubric of having decided to follow a given policy, where 'policy' has some rather specific implications.

For what reasons do we accept the conclusions of arguments which are probabilistic? Needless to say this constitutes a problem for any theory of justification. Sellars urges reconisdering the meaning of 'probable' and 'probability'. He suggests that we start from the common sense meaning of 'probable' as 'there are good reasons'. Using what in many ways resembles a Carnapian 'logical theory of probability', he develops this theme by considering different types of good reasons. This results in a proliferation of distinctions. There turn out to be at least three different *modes* of probability. Furthermore, with respect to any given probability argument there are three different types of 'outcomes'. But despite the complexity of the theory two major points stand out: (a) Sellars' inductive rule of acceptance for empirical generalizations is the straight rule; (b) the claim that we should use a straight rule of inductive acceptance in probability arguments becomes a premise in a practical argument with the conclusion that we should in fact accept that-p, where 'that-p' constitutes the conclusion proper of a probability argument.

The crucial problem with this account arises in terms of the reasons for using the straight rule. The drift of Sellars' argument is that the policy of continued acceptance of descriptions which are 'probable' puts us in the best possible condition to accept *frameworks* which are 'probable', and that the obtaining of such frameworks is the objective of scientific inquiry. Having such frameworks puts us in the position whereby we can "give non-trivial explanatory accounts of extablished laws". ([37], p. 198) ('Laws' here refers to empirical generalizations.) We accept statements which are probable on the hope that doing so will lead to a framework which produces good descriptions, explanations and predictions.

Between the stages of adopting theories and generalizations lies a point at which we adopt rules of inference. Sellars deems it reasonable to espouse a principle of inference if no evidence indicates its espousal will produce falsehoods. Postulating an order of dependence among the different modes of probability. Sellars argues that the probability of theories depends on the probability of rules of inferences (laws of nature). In discussing the probability of rules of inference, what he calls 'nomological probability', he makes a number of interesting points about the relation between evidence and lawlike statements. First, claiming a given proposition is probable is not to assert that it stands in a given relation to a certain body of evidence. It is rather, to report that the statement has good reasons for being adopted. In this sense, then, a probability claim is an appraisal. Second, we accept rules of inference, not empirical claims, when we adopt the conclusion of a nomological probability argument. That is, we do not adopt the conclusions of the probability argument and then proceed to make predictions and offer explanations using it. Rather, we use the conclusion of probability arguments in the postulation of rules of inference.

To rationally accept the conclusion of the probability argument on traditional grounds requires a rule of acceptance. But Sellars does not believe that there are any rules of acceptance for non-deductive arguments. As noted in Chapter II, given Sellars' account of logic, to call a method of reasoning logical means that there is a criterion of soundness for those arguments. Thus, if an argument is sound, there is no reason not to accept its conclusion, since it is true. Implicit in this account is the idea that the only viable rule of acceptance is one that says we ought to accept true statements. To the extent that arguments cannot be guaranteed to produce true conclusions, given true

premises, then their conclusions should not be accepted as true. If we weaken
the requirement that the conclusion be true to require merely that it have a
high probability, then the problem arises as to how high a probability is
sufficient. And, because we can show that any number less than one on a
probability scale is insufficient to warrant the acceptability of the argument,
we are again forced to abandon the idea that the conclusion *per se* of a probability argument is acceptable. A typical example used to illustrate the
insufficiency of a high probability to warrant acceptance is the lottery
paradox. Salmon gives a rather straightforward account of this problem.

If inductive logic contains rules of inference which enable us to draw conclusions from
premises – much as in deductive logic – then there is presumable some number r which
constitutes a lower bound for acceptance. Accordingly, any hypothesis h whose probability on the total available relevant evidence is greater than or equal to r can be accepted
on the basis of that evidence. (Of course, h might subsequently have to be rejected on
the basis of further evidence.) The problem is to select an appropriate value for r. It
seems that no value is satisfactory, for no matter how large r is, provided it is less than
one, we can construct a fair lottery with a sufficient number of tickets to be able to
say for each ticket that it will not win, because the probability of its not winning is
greater than r. From this we can conclude that no ticket will win, which contradicts
the stipulation that this is a fair lottery – no lottery can be considered fair if there is
no winning ticket. ([37], p. 221)

To avoid this situation Sellars claims that we do not *accept* the conclusions
of probability arguments. Rather, we *use* them to formulate rules of inference.
The rule of inference notes that given that a certain state of affairs occurs
under a given set of circumstances x per cent of the time, we ought to infer
the probability that there is an x chance of that state of affairs occurring
the next time that set of circumstances obtains. Taking this line allegedly
does not lead to the lottery paradox.

In his discussion of nomological probability Sellars also tackles the question raised in Chapter II concerning which material rules of inference to
accept. Accept those which represent the facts. Our ordinary day-to-day
inferences can be vindicated if, on the basis of these inferences, we come to
develop laws and theories which do not permit us to make unsound inferences.
This, in turn, says that what works is what we have and if we have it then it's
justified, all of which in turn sounds very much like our reading of Goodman.
But, then again, this begs the issue of justification.

A theory of justification tells us what we ought to do in order to give good
reasons for accepting or rejecting theories, laws, etc. But to say that 'you
ought to act in accordance with the facts' seems to miss the point, since,

to paraphrase Goodman, and keeping in the Kantian spirit, the facts are where you find them depending on the tools, i.e., theories, you are using. It is, at best, only a ncessary and not a sufficient condition for accepting a rule of inference that the world has in the past in fact looked the way the rule says it should. To put this in slightly different terms, Sellars does not manage to counter Hume and Goodman's classic observation that even if things have appeared ever so regular in the past we cannot assume they will continue to be so regular in the future.

4. THE MEANING OF 'PROBABLE'

In the basic non-metrical sense of 'probable' (in relation which all other senses are to be understood), to say of a statement or proposition that it is probable is, in first approximation, to say that it is worthy of credence, that is, to put it in a way which points toward a finer grained analysis, it is to say that relevant things considered there is good reason to accept it. ([42], p. 198)

In the finer grained analysis to which Sellars alludes he distinguishes two different types of arguments and three different outcomes of these arguments. The state of accepting the proposition in question constitutes the 'terminal outcome'. The remaining two outcomes, the proximate and the practical, are the respective proper conclusions of a probability argument and a practical argument.

Statements are probable. But there are different kinds of statements, e.g., laws and theories (which in this context Sellars views as a conjunction of the principles of a theory). With respect to each of these kinds there corresponds a mode of probability. There is a different set of good reasons, or a different type of reasoning by which the good reasons are supplied, for accepting different types of statements.

A probability statement is the conclusion of a probability argument; and,

there are clearly such things as 'probability arguments' where this simply means an argument of which the conclusion is a probability statement, i.e., a statement to the effect that a certain other statement or proposition is probable. ([42], p. 199)

At this point, however, we must be careful. The mere recognition of probability arguments does not commit Sellars to treating probability in anything like a traditional manner. And he takes pains to offer a warning on just this point. To admit probability arguments, "involves no commitment concerning either the form of the argument or the form of the conclusion". ([42], p. 199) A similar warning emerges when he first introduces his three outcomes:

I distinguish between the 'proximate', the 'practical' and the 'terminal' outcomes of what I shall call first-order probability arguments. My reason for using the expression 'outcome' rather than 'conclusion' (which might seem a more appropriate way of characterizing the *terminus ad quem* of an argument of inference) is that although the whole point of the first order probability argument is to generate a terminal outcome, the relation between a terminal outcome, and the premises of the argument is radically different from that which obtains between what we ordinarily call the conclusion of an argument and its premises. ([42], p. 199)

A practical outcome constitutes the *terminus ad quem* of a practical argument and a proximate outcome the *terminus ad quem* of a probability argument. The relation between these two types of argument is complicated, but essential to Sellars' entire program.

15. That there must be a practical outcome of a first order probability argument is an immediate consequence of our tentative analysis of probability statements. For if

it is probable that-p

has anything like the sense of

relevant things considered, it is reasonable to accept that-p

then it tells us that

there is a good argument which takes relevant things into account and has as its conclusion *I shall accept that-p*

that is to say, it points toward an argument which can be schematically represented as

. . . .
therefore, I shall accept that-p

16. Reflection on the above makes it clear that on the view I am proposing there are at least two arguments in the logical neighborhood of every probability statement.

(a)   the first order probability reasoning, yet to be explored, which can be schematized

. . . .
therefore, it is probable that-p

(b)   the practical reasoning schematized above as

. . . .
therefore, I shall accept that-p

the important thing to note is that of these two reasonings, the former, although it is what I am calling a first order probability argument, stands to the latter, roughly, as meta-argument to object-argument. More accurately, the conclusion of the former asserts that the premises are available for a good and valid argument of one or other of a number of patterns of which the conclusion is the conclusion of the reasoning schematized in (b). ([42], p. 201)

The probability argument stands to the practical argument as meta-argument to object-argument. Thus, some of the premises used in the probability argument can be used in the practical argument. But there is more involved here than meets the eye. The practical argument is itself *suggested* by the conclusion of the probability argument (hence, indirectly by the entire argument) because the conclusion, being a probability statement, says there is such a practical argument which can be constructed. But such a view leads to problems. Specifically, the use of a practical argument leading to the conclusion

> therefore, I shall accept that-p

is unnecessary. To begin with, Sellars argues that where there is a probability argument there is a practical argument in its logical neighborhood leading to the conclusion that the conclusion of the probability argument should be accepted as true. This seems to be the case for any probability argument. If so, why bother with all the complications of logical neighborhoods? Let us merely institute an urban renewal project and sweep away the slums leaving simple probability arguments. Since we now all know the meaning of 'probable' is 'there are good reasons to accept', then whenever we can produce a probability argument we can accept its conclusion. Thus we can accept anything at all since we can always produce an argument giving a certain conclusion some probability or other.

At this point it is not enough to counter that Sellars claims not merely that there is an argument with a practical outcome, but there are *good* reasons for accepting the conclusion. For this merely begs the issue. You've got to take Sellars' whole view or reject the lot it seems. For Sellars' only way out here is to pursue the line that the very purpose of science is to produce better explanations and that you need to be guided not by a sloppy notion of 'good' reasons which could change as the direction of wind does as in a Jamesian Pragmatism, but by the end in view, a Peircean final theory.

I think Sellars is sensitive to this point for he argues that it is not possible to construct a practical argument on the basis of a probability argument in all cases. The constructability of a practical argument is not obvious from the schematic account of a first order probability argument because it is *incomplete.*

. . . the real reason why

> it is probable that-p

is 'incomplete' is that it points to a practical argument which it doesn't enable one to construct. Moreover, the reason why what I shall call the 'second order probability statement'

it is probable that-p in relation to $e \cdot R_L$ (that-p, e)

is more satisfactory is not that it is as a whole a non-elliptical probability statement – for it is not, having the form

$e \cdot R_L$ (that-p, e) implies that it is probable that-p,

the reason is, *that it enables the construction of certain crucial features of the practical argument* in question. Thus it enables construction of practical arguments of the following form

.
.
.

q
R_L (that-p, that-q)

.
.
.

therefore, I shall accept that-p ([42], p. 203)

The second order probability statement is *complete* because it allows us to construct key features of the corresponding practical argument. These key features are the evidence for the claim that-p and the relation that obtains between that-p and its evidence. In the example given 'q' denote the evidence, 'that-p' the statement asserting the evidence, and 'R_L' the logical relation claimed to hold between q and that-p.

In the schematic version of a second order probability statement noted above, 'e' refers to the total relevant evidence. Sellars provides no criterion of relevancy, but it is not a crucial point for this discussion. Needless to say, spelling out the relevancy requirement raises a troublesome issue wherever it appears. Speaking to the total relevant evidence question in a footnote, Sellars notes:

In these formulations from now on I shall leave implicit the requirement that e be the total relevant evidence or grounds on which that-p is probable. I shall absorb this requirement into the symbol 'R_L' in the sense that (a) I shall limit my discussion to pure cases of the different modes of probability, there being only one R_L in each pure case; (b) I shall assume that in each pure case R_L obtains between that-p and all the evidence (or, more generally, grounds) appropriate to the particular example of that pure mode. ([42], p. 203)

Thus, while there is a logical relation between the evidence and the statement in question, 'probable' does *not* name that relation. This does not entail that 'probable' does not name any logical relation. In the context of 'it is probable that-p', 'probable' means:

> there is a good argument of one or other of certain patterns for accepting that-p. ([42], p. 202)

The distinction between different levels of probability arguments at this point serves as a way of marking out those probability arguments which are complete from probability statements where the concept occurs in the non-metrical sense originally noted. But if we look a little closer we can see that there is no argument here defending the claim that a first order probability argument is essentially incomplete and that it is reasonable to continue to support the levels view of probability. There *is* an account of the completeness of the second order probability argument. It rests on the claim that the second order probability argument supplies the key ingredient permitting the construction of a practical argument leading to the acceptance of that-p. This item is 'R_L'. That is, a second order probability argument is complete because it mentions the logical relationship holding between e and that-p. Mention of this relation permits constructing a practical argument leading to the conclusion

> I shall accept that-p

because, given it, we know what the major premise of the practical argument must be:

> Accepting conclusions of those arguments establishing R_L between that-q and that-p.

Now since no such item as 'R_L' appears in a first order probability argument we can't construct a practical argument leading to the acceptance of its conclusion. This leaves us to conclude either that first order probability arguments aren't really probability arguments or that Sellars has no legitimate way out of concluding that we can accept the conclusion of any first order probability argument.

We are forced to conclude the former if we take Sellars' account of 'probable' seriously. To claim that a statement is probable is to claim there are good reasons for accepting it. Since for first order probability arguments there are no good reasons available for accepting their concluctions we seem complelled to conclude that they must not be probability arguments. But, it might be argued, there is a difference between claiming that there are good reasons for accepting a conclusion and that there is an argument with the conclusion that you should accept a certain conclusion. Indeed there is such a difference, but it doesn't help Sellars here. For by asserting that a first order probability argument is *essentially* incomplete he is effectively claiming that no practical argument can be constructed, i.e., there are no good reasons for accepting that-p. If that is so then how can

it is probable that-p

be a meaningful claim?

On the other hand, Sellars is forced to admit that we can accept the conclusion of any first order probability argument since by definition there are good reasons for accepting the conclusion, even though we can't provde them. It is apparent that this too is an unacceptable conclusion.

5. 'PROBABLE' VERSUS THE GROUND-CONSEQUENCE RELATION

Rephrasing the results of the previous section: Sellars wants to view probability not as a relation but an appraisal. In a sense, however, this is too strong. While claiming there are good reasons to accept a claim is to appraise it, when we say of a statement that it is probable we also recognize a relation between that statement and something else. According to Sellars, the 'something else' denotes the practical argument which results in accepting that-p and *not* the evidence for the statement.

Sellars' argument for considering probability as a special relation consists of three parts. First, he suggests that like 'father', not every relational notion need mention that which it relates. This observation serves as the basis for his discussion of why probability should not be characterized as a logical relation between a statement and its evidence. Armed with these preliminaries, and his prior analysis of the meaning of 'probable', Sellars continues to the central point of his theory, that probability arguments are not inductive arguments simply put. This can be rephrased in a Sellarsian fashion by saying that the relationship between the premises and the conclusion in an inductive argument differs from, and should not be confused with, the relationship which obtains between a probability argument and its conclusion. The conclusion of an inductive argument is a statement which may or may not be true. The outcome of a probability argument, properly conceived, is the *actual* acceptance of the claim that-p, hence, different in kind. All of this, however, does not mitigate the conclusion that Sellars cannot specify why we should use 'probable' as a relation in first order probability arguments since there is no practical argument to which it points.

Turning to the first consideration of the possibility that probability might be a relation, we are confronted by a truly first class piece of Sellarsian dialectical reasoning. Starting from his claim that statements are probable, Sellars notes this assertion might lead someone to interpret him as making probability an intrinsic property of statements. He denies he intends this. Proceeding then on the course which an imaginary critic might take: if probability is not an intrinsic quality of some statement, then it must be a relation. And now we are on the dialectical road.

Sellars first calls this last move a *non-sequitur*, if what the critic means by 'relation' amounts to something like simple two-place predicates.

If we put the above by saying that probability is a 'property of propositions', we run the danger of being interpreted to mean that probability is an 'intrinsic' characteristic of propositions, something they could have regardless of what else was the case. Since the latter is obviously absurd, the conclusion is often drawn that probability must be a *relation*. That this is a non-sequitur appears immediately one takes into account the existence of generalized relational properties. Thus compare

(1)   Tom begot John
(2)   (*Ex*) Tom begot *x*
(3)   Tom is a father

it would be an egregious blunder to say that 'father' in (3) is a relation word, a two-place predicate. Yet it would be equally mistaken to infer from this that Tom's being a father is logically independent of the existence of something to which Tom stands in the relation expressed by ' ... begot ... ' or ' ... is the father of ... ' But more of this later when we attempt to determine just what kind of relational property probability is. ([42], pp. 198–199)

Probability is a relation of *some* kind; now to determine its nature.

Sellars claims we can deny probability is an intrinsic property of propositions and still not be forced to characterize it as a relation. But he has not thereby countered the claim that probability is *nothing more than* a relation. A probability statement here becomes merely a shorthand account of a statement which specifies the evidence for the claim and the relation in which the claim stands in that evidence.

Insisting that a full explication of 'probable' requires articulating the evidence, *e*, for a claim, *p*, and the relation between *e* and *p*, leads, he says, to a vicious circularity. If '*p* is probable' specifies a relation between *p* and its evidence, then the evidence *implies p* to a certain degree. And if we ask what 'implies' means in this context, that is, if we ask for the logic of inductive inference, all we get is: 'stands in a given relation to . . . '.

It is sometimes said that the fact that probability is relative to evidence requires that the conclusion of a probability argument have the form

> it is probable that-p in relation to total evidence e such that R_L (that-p, e)

and that

> it is probable that-p

makes sense only as an 'elliptical' formulation of the more complicated statement. This is obviously false if intended to mean that a well formed probability statement must mention specific evidence and specify the logical relation in which the proposition in question stands to the evidence. On the other hand, it is indeed true that 'it is probable that-p' is in some sense elliptical for a statement making a general reference to the evidence and its relation to the proposition that-p. But not, and this is the crucial point, for a statement of the form

> some state of affairs obtains such that it is probable that-p in relation to it

For if we ask 'what is meant by "in relation to" '? the answer, if our analysis is correct, must be *implies*, so that the above is equivalent to

> some state of affairs obtains which *implies* that it is probable that-p

and the expression of which the 'ellipticality' was in question recurs. ([42], p. 202)

But, Sellars has already said probability *is* a relational property. If it does not establish this specific relation, to what type of relation does it pertain? Recall that Sellars calls 'father', as it occurs in (3), a generalized relational property. 'Probable' also stands for a general relational property. This general property refers generally to evidence and its relation to the statement in question. It does this task of general referring not by mentioning specific evidence but by implying there exists an argument which has as its conclusion,

> I shall accept that-p.

It entails there are good reasons, i.e., an argument, for accepting that-p, without specifying what those reasons are. Thus,

the solution of the puzzle emerges from our account of

> it is probable that-p

as equivalent to

> there is a good argument of one or other of certain patterns for accepting that-p

According to this analysis a probability statement is 'elliptical' for a statement making a *general* reference both to evidence and the relation to the proposition that-*p*; but the statement for which a probability statement is thus 'elliptical' is not, itself, a 'non-elliptical' statement involving the predicate 'probable' in two-place glory, but simply its *analysis*. ([42], p. 202)

But, if the conclusion of a probability argument does not refer to its factual evidence, but to the 'practical evidence', i.e., the practical argument, for its acceptance, then how are we to read the conclusions of probability arguments? For one thing, Sellars argues, we are to read them as conclusions of inductive arguments, where we would determine the argument valid in accordance with the rules of an inductive logic.

If my thesis, *positively* put, urges a distinction between two reasonings in the neighborhood of a probability statement (each of which, we shall see, is deductive when made fully explicit), then it can also be *negatively* put by urging that the approach to probability through the concept of non-demonstrative inferences — 'valid after their own kind' — of the form

$$\frac{e}{\text{therefore } h}$$

or perhaps

$$\frac{e \cdot R_L \ (h, e)}{\text{therefore } h}$$

or even

$$\frac{e \cdot R_L(h, e)}{\text{therefore probably, } h}$$

is, as I see it, to confuse the relation between the terminal outcome of an argument pertaining to probability and its premises with the ground-consequence relation. ([42], p. 202)

Probability reasoning is not a special form of inductive reasoning. '*P* is probable' does not mean we have a certain amount of factual evidence which we can relate to *p* by the rules of some non-deductive logic. Rather, '*p* is probable' means there is a good argument for accepting what *p* says as true, except, as we have seen, in the case of first order probability arguments.

Sellars does not intend this to imply there is no such thing as inductive reasoning. Rather he objects to using the probability calculus to reason from inductively obtained data to a fast and firm conclusion *in one fell swoop*.

Probability reasoning entitles one to conclude that-p is true, or ought to be believed to be true, or accepted as true. No one set of rules entitles you to go directly from inductively obtained data to the acceptance of p. The acceptance of p remains a psychological phenomenon, the terminal, i.e., final, outcome of the use of probability. On the other hand, the acceptance of that-p, to be rational, must be reasoned. Sellars locates reasoning which licenses accepting p in a *practical* argument with the conclusion

I shall accept that-p.

This constitutes the practical outcome of the probability reasoning. It follows *deductively* from a set of premises just as the proximate outcome, i.e., the conclusion, of the probability argument does. The analysis of the conclusion of the probability argument comes to: there are good reasons to accept the conclusion of this argument, which reasons are to be found in another argument, one which has as its conclusion that

I shall accept that-p.

6. THE PURPOSE OF PROBABILITY ARGUMENTS

Simply put, the creation of an explanatory framework constitutes the goal of all scientific and probability reasoning. The conceptual ability to draw inferences constitutes a logically necessary condition of being in an explanatory framework, F_e. This requires rules of reference. But, while rules of inference are a logical prerequisite to theoretical explanation, Sellars insists the form of probability reasoning stands out clearest with respect to the probability of theories. Even though the structure of probability stands out clearest with respect to theories, the *probability* of theories *rests* on nomological probability. Nomological probability, in turn, rests on the form of reasoning to which *it* is logically prior,[1] namely the form of statistical probability involved in the compiling of data on the basis of sampling population samples. In short, "there is a logical order of dependence among the various modes of probability". ([42], p. 220)

We use the mode of probability relating to theories to put us in a position where, with the appropriate practical argument, we can accept the theory in question. With respect to the mode of probability resulting in the acceptance of rules of inference, i.e., nomological probability, its use put us in a position to argue for a given theoretical account.

For the end-in-view in *nomological* induction (which must not be confused with the mode of induction involved in the statistical or proportional syllogism to be examined in a moment) is not the possession of empirical truth, but the realizing of a logically necessary condition of being in the very framework of explanation and prediction, i.e. being able to draw inferences concerning the unknown and give explanatory accounts of the known. ([42], p. 219)

When one argues on the basis of a rule of inference resulting from a probability argument one operates within F_e, the framework of explanation. A rule of inference, the product of a complicated process is not an empirical claim. This process involves a practical argument leading to the conclusion that one should accept the rule of inference given the results of a logically prior probability argument, the conclusion of which is an empirical claim, i.e., the result of an argument of the statistical or proportional mode of probability. We therefore formulate the rules of inference on the basis of inadequate data, i.e., less than complete. Thus the possibility exists that the rule of inference resulting from these prior calculations and patterns of reasoning may be false. A false rule of inference is one we assume to be true and yet leads one to false conclusions.

A principle of inference, whether a logical principle in the narrower sense or a physical principle of inference (*P*-implication), is known to be false if an argument made in accordance with it is known to have a true premise but a false conclusion. Suppose, now, that at a time t, when 3/4 of the observed cases of A have been found to be B, I reason myself into accepting the principle of inference that

K_1 is a finite unexamined class of As *P*-implies that approximately 3/4 ΔK_1 is B

and then proceed, for a specified ΔK, ΔK_1, to infer

(1) ΔK_1 is a finite unexamined class of As therefore approximately 3/4 ΔK_1 is B.

Suppose, however, it turns out, subsequently, at t', that only 1/4 ΔK_1 is B. Then, of course, the premise of (1) was true, but the conclusion false. Consequently my inductive reasoning led me to espouse a false principle of inference. But, and this is a logical principle, inductive reasoning as we have analyzed it, can never lead me to espouse a principle which I know to be false at the time of espousal. For until ΔK_1 is examined, the principle is not known to be false, and after ΔK_1 is examined, the principle no longer applies to ΔK_1. The inductive reasoning which is (at, t') relevant will lead me to espouse a modified principle. This principle, in turn, may turn out to be false, but this fact in no way impugns the rationality of the inductive enterprise. ([42], pp. 218–219)

We do not impugn the rationality of the inductive enterprise since it proceeds on the basis of true premises, e.g., that as a matter of fact at the time, t', the sample was in fact such and such. If the new sample examined at t'' differs proportionally from the sample examined at t', that still does not falsify the truth of the claim made at t'. Thus, it is rational to infer that the future will be like the past, if in fact what you know about the past happens to be true. As it stands this hardly seems reasonable, as Goodman has pointed out.

Inductive reasoning, according to Sellars, is a perfectly good method of reasoning provided it uses true premises. Moreover, both true premises and rules of inference remain the objective of using probability arguments. We formulate empirical claims on the grounds of a given amount of information. Then, on the basis of these empirical claims, we argue probabilistically for rules of inference using a higher mode of reasoning. The conjunction of a given set of these rules forms the basis for F_e, the test of which is whether or not we can make correct inductive inferences concerning unknown phenomena. For Sellars, our ability to make correct inductive inferences at the level of proportional or statistical probability, based on inductive reasoning in the first place, vindicates the policy to use inductive generalization in this way.

Probability arguments are to provide us with the framework to make the inductive inferences which are predictions and to offer explanations. These arguments clearly differ from inductive inferences in a plebian sense. The conclusion proper of a full probability argument, according to Sellars, is the acceptance of a claim as true, while an empirical claim results from normal inductive inferences. When we test inductively generated empirical claims we are testing the rule of inference which permitted the proposing of the claim in the first place. This provides another check on our policy to use induction.

7. PRACTICAL REASONING

Given Sellars' version of the meaning of 'probable', however inadequate it turns out to be, we can at least understand the claim that "at the heart of the concept of probability is the concept of a form of practical reasoning". ([42], p. 204) If probable means anything like there are good reasons to accept, and if, as Sellars reasons, the giving of good reasons occurs in a practical argument, then we are led naturally to the notion of practical reasoning.

I intend to only outline some of the more salient features of Sellars' theory of practical reason. In particular, I consider only those parts essential to his theory of probability since I am not here concerned with the concept of practical reasoning *per se*. Despite Sellars' claim that practical reasoning constitutes the core of a probability argument, the crucial problems still lie in that part of the theory of probability which produces a statement of the form

> *h* is probable.

Sellars' theory of practical reason comes in three parts: a succinct statement of the key features of his theory, the logic of the theory, and a distinction between the types of inference.

(1) Let us say that utterances of the kind illustrated by 'I shall do *A*' and 'It shall be the case that-*p*' express INTENTIONS, as utterances of the kind illustrated by 'it is raining' express statements or propositions. Note that the intention expressed by 'Tom shall make amends' is not Tom's intention, but the intention of the speaker.

(2) One particularly important variety of intention is the conditional contention, thus

(i) I shall do *A*, if *p*
(ii) Tom shall make amends, if he is guilty.

Note that the condition 'if *p*' is intrinsic to the intention. This can be brought out by reformulating these intentions as

(iii) Shall (my doing *A*, if *p*)
(iv) Shall (Tom making amends, if he is guilty)

(3) A particularly important variety of conditional intention for our purposes is the general conditional intention

(v) I shall do *A* whenever *X* obtains.

A GENERAL conditional intention can be called a policy. ([42], p. 204)

The following simple logical principle governs the application of these ideas:

> If (. . .) implies (- - -)

then

> [shall (. . .)] implies [shall (- - -)].

To explicate both 'implies' and the above mentioned distinction between different types of inferences, consider again Sellars' own statement.

Complications arise only when we take into account, as we must, the difference between primary and dependent implication. Let me explain.

(a)   By 'implication' I mean a relation (between propositions or intentions) which authorizes inference.

(b)   By calling the implication between P and Q a *dependent* implication, I mean that it presupposes the truth of an unmentioned proposition R (which is, however, indentified in one way or another by the context) and which is such that the less dependent implication

$P \cdot R$ implies Q

obtains.

(c)   If we represent an independent implication by

$\alpha \cdot \beta \to \phi$

Then the dependent implication which holds between α and ϕ, presupposing the truth of β, can be represented

$$\alpha \overset{\beta}{\to} \phi$$

It was pointed out above that

P implies Q

presupposes the truth of P and correspondingly implies the truth of Q. The same situation obtains with respect to dependent implication, thus the dependent arrow statement pictured above presupposes the truth of the explicitly mentioned α and of the contextually identifiable β, and consequestly implies the truth of ϕ. ([42], p. 205)

The theory of practical reason which emerges from these points helps clarify the reasoning surrounding Sellars' concept of probability. Given the following typical practical argument:

I shall accept a proposition if it satisfies condition C

h satisfies condition C
Therefore I shall accept h

We can write it as a non-dependent implication:

'I shall accept a proposition if it satisfies C and h satisfies C' implies 'I shall accept h'.

We also can diagram the argument.

I shall accept a proposition if it
satisfies C
h satisfies C ⌐⎯⎯⎯⎯→ 'I shall accept h'

The diagram exhibits the dependent implication between 'h satisfies C'

and 'I shall accept h', and the presupposition concerning acceptance of a proposition embedded in the major premise. Probability arguments and practical reasoning share a structural feature. Both

h is probable

and 'I shall accept h' are consequents of dependent implications. To see this fully requires an analysis of probability reasoning in probability arguments *per se*. That is the subject of the next section. For present purposes a brief account nevertheless is still possible.

It has already been pointed out that 'makes', in this context is equivalent to 'implies'. Here again we have a dependent implication; only this time, the implication is the principle of a dependently valid probability argument as contrasted with the dependently valid practical argument dicussed above. What, then, is the valid argument, with e and $R_L(h, e)$ as premises, which has 'h is probable' as its conclusion? The answer is not far to seek, it must be something like

> a proposition is probable if it stands in $R_L(h, e)$
> e is the case
> $\therefore h$ is probable$_M$

The desired dependent implication is

$$\underline{e} \cdot \underline{R_L}(h, e) \quad \overset{\text{\Large\textbar}\begin{array}{l} \textit{a proposition is probable}_M \textit{ if it stands} \\ \textit{in } R_L \textit{ to a fact} \end{array}}{\longrightarrow} \textit{'h is probable}_M\textit{'} \ ([42], \text{p. 207})$$

The underlining of the phrase on the left hand side of the arrow, the antecedent of the dependent conditional, indicates that it is posited as true.

With the form of the probability argument outlined we come to the crucial point. The principle of the dependent argument is: 'a proposition is probable$_M$ if it satisfies condition C'. To say a proposition or statement is probable means there are good reasons for accepting it. This amounts to:

> there is a good argument of kind M for accepting a proposition if it satisfies condition C. ([42], p. 207)

That is, provided that the proposition in question meets certain conditions, we can construct an argument providing the reason for accepting the proposition. As Sellars notes,

there is a good argument of kind M which has as its conclusion 'I shall accept a proposition, if it satisfies condition C' ([42], p. 207)

There are two further items required to clarify the rest of Sellars' discussion.

First, the condition C is not an arbitrary limit set on the argument. The condition is designed with respect to a given goal, the purpose behind using the argument in the first place. We pursue a given policy in the active attempt to achieve this goal. A policy is a general conditional intention. Thus to complete the argument we need some statement of the policy which legitimizes the formulation of condition C. Second, the statement of policy forms the major premise in the complete practical argument. It is also crucial to our discussion of the modes of probability because, according to Sellars, we use different policies within the different modes of probability arguments.

In short, the major premise of the first order probability$_M$ argument tells us that the practical reasoning which culminates in

I shall accept h

(where this acceptance is bound up with probability$_M$) has the form

I shall bring about E

(but bringing about E implies accepting a proposition, if it satisfies condition C)
So, I shall accept a proposition, if it satisfies condition C

h satisfies condition C

so, I shall accept h. ([42], pp. 207–208)

Now to locate the position of the probability argument within the general practical argument. Let us first reconsider the similarity in question between

I shall accept h

and

h is probable$_M$.

They are both consequents of dependent implications. As such they point to the principles which make their respective arguments valid. In the practical argument, the principle in question is

I shall accept h if it meets condition C.

In the probability argument, the principle is

A proposition is probable$_M$ if it stands in R_L to the relevant facts.

The probability argument culminating in 'h is probable$_M$' constitutes the reason for saying 'h satisfies condition C'. This statement is one of the premises in the argument culminating in 'I shall accept h'. Thus, the probability argument occurs in the context of saying that h as a matter of fact does meet condition C.

We have already emphasized that a first level probability$_M$ statement is a statement to the effect that there is available a good piece of practical reasoning of a certain kind which has as its conclusion 'I shall accept h'. It should now be clear that a first level probability argument . . . is a meta-practical argument which *establishes* the availability of this piece of practical reasoning. It does so by deriving the existence of a good argument for accepting the specific hypothesis in question from the existence of a good argument for accepting any hypothesis which satisfies the condition which the given hypothesis is known to satisfy. ([42], p. 207)

Now, however, the initial plausibility of Sellars' claim that 'probable' means 'there are good reasons' fades. Take the following as a concise statement of the analysis of 'probable':

Thus

h is probable$_M$

where the subscript indicates a specific mode of probability, asserts the availability of a good argument for 'I shall accept h', the ultimate major of which is the intention to achieve a certain end, and the proximate major is the appropriate intention to follow a certain policy with respect to accepting propositions. ([42], p. 208)

This is surely misleading if it entails that the practical argument in question will not be available unless, as a matter of fact, the *policy* which permits its formulation is operative. For now we would have to weaken Sellars' analysis of 'probable' as there are good reasons. We could only conclude that *if* there is a policy which urges accepting statements meeting conditions of kind C, *then*

h is probable

suggests that a practical argument leading to the conclusion that

I should accept h

can be constructed.

This objection, however, fails. The policy we claim is not necessarily present is indeed present. Recall the purpose behind probability arguments: to provide the necessary logical equipment for F_e. In F_e we can explain phenomena by telling what they are and how they behave. We accomplish this by using rules of inference. F_e remains our goal. Pursuant to that objective we implement subsidiary policies we believe will bring us there. Such a subsidiary policy is the premise in the complete practical argument cited above. This premise Sellars calls the proximate major:

I shall accept a proposition, if it satisfies condition C.

Note that, according to Sellars, the general conditional intention that-(I shall bring about E), the ultimate major, entails this subpolicy.

Let us also recall that the modes of probability stand in an order of logical dependence. This too will have an important bearing on the sense in which

h is probable$_M$

does indeed *have* a practical argument in its logical neighborhood. But this order of dependence in which modes of probability stand creates its own problems. For the fact remains that if you introduce levels to bring order out of chaos you cannot ignore them when they prove difficult. In this the difficulty arises out of the problem raised at the end of the last section, namely that there are severe problems with Sellars' account of 'probable' for first order probability arguments. If the general conditional policy entails subpolicies why does this process of entailment not extend down to the level of first order probability arguments? Perhaps the answer lies in the recognition that to do so is extremely difficult, in fact impossible, if one wants to avoid the lottery paradox. For the condition C that a first order probability claim must meet seems to be a high enough probability to merit acceptance on no other grounds. Awareness of this requirement might be the reason why Sellars claims that the conclusion of a first order probability argument is non-metric. That way he avoids the lottery paradox by ignoring it.

But that gambit won't work either. For the first order probability argument conclusion acts as a promissory note for a practical argument which cannot be completed, but which is a logical consequence of the general policy behind the use of probability in the first place. Thus we are forced to conclude either that there is no good reason for accepting the conclusion of a first order probability argument or that the reason for accepting it is the result of a series of implied policies stemming from a general policy, which policy in turn is a result of the basic non-metric sense of 'probable'. This last approach is clearly circular.

8. MODES OF PROBABILITY

Sellars not only believes that practical reasoning lies at the heart of probability, he also remains convinced of two other points about probability, both of which have a bearing on the structure of arguments leading to the conclusion

h is probable.

First, depending on the nature of h the mode of probability varies. h can be either a theory, a rule of inference or a statistical empirical claim. For Sellars there are different modes of probability; the mode of probability is a function of the character of h.

We have already taken note of two different types of probability arguments: first and second order. They differ in that the conclusion of a second order probability argument allows us to reconstruct certain key features of its corresponding practical argument while the first order probability argument is incomplete in this respect. As Sellars puts it,

. . . the first order probability statement

It is probable$_M$ that-p

where the subscript 'M' indicates a specific mode of probability, serves as a promissory note authorization of

I shall accept that-p. ([42], pp. 206–207)

It remains a *promissory note* because the argument it points to cannot be complete; we lack the relevant information. A second order probability statement does provide the relevant information. A second order probability statement permits the reconstruction of the relevant practical argument because it mentions the reasons for 'h is probable'. In analyzing this type of statement the sense of which 'h is probable' and 'I shall accept h' are consequents of dependent implications becomes clear.

$e \cdot R_L \, (h, \, e)$ makes h probable

Secondly there is a logical order of dependence among the different modes of probability. The availability of a conclusion

h is probable

where h is a theory, depends on the prior availability of conclusions of which 'h' is a rule of inference. This in turn depends on the prior availability of statistical claims derived from arguments of the appropriate mode of probability. These, as it turns out, depend on a prior use of inductive reasoning.

In this section I follow Sellars' procedure and concentrate on that mode of probability appropriate to theories. But, as becomes increasingly apparent, while the structure of probability arguments with respect to theories may stand out most clearly, the interesting issues surround the mode of probability which produces statistical claims. For at this level we first encounter the concept of probability in metrical guise. The immediate problem then is to see why Sellars' use of the straight rule should not lead him into all the problems traditionally associated with this approach to metrical probability, e.g., Goodman's true problems. The answer lies in Sellars' use of the straight rule to produce only empirical claims. The *use* of those empirical claims requires rules of inference, the adoption of which is not an empirical matter at all, but a practical one. But since this rests on the first order probability argument, it appears that we can't get off the ground.

35. The mode of probability in connection with which the relevant factors stand out most clearly is that of the *probability of theories* – in spite of the fact that the latter logically involves all the other forms of probability. Consider the following piece of reasoning, where 'T' is a surrogate for the sentence which expresses the conjunctive proposition the conjuncts of which are the principles of a certain theory.

> T is the simplest available framework which generates acceptable approximations of nomologically probable lawlike statements and generates no falsified lawlike statements [from now on, the statement that T has this complex property, will be represented by '$\phi(T)$']

Therefore, I shall accept T.

36. Notice that this is (a) a practical argument; (b) an enthymene. Its principle is, therefore, a dependent implication and, indeed

$$\text{'}\phi(T)\text{'} \quad \begin{array}{|l} I \text{ shall accept frameworks which are } \phi \\ \hline \longrightarrow \text{'I shall accept } T\text{'} \end{array}$$

To this implication there corresponds the probability$_t$ statement

> T is probable$_t$

i.e. the meta-practical statement

> there is something which is the case which dependently implies 'I shall accept T'

and correspondingly to this the meta-probability$_t$ statement

> $\phi(T)$ makes it probable$_t$ that T is true

i.e.

$$\phi(T) \quad \begin{array}{|l} \cdot \;\; \cdot \;\; \cdot \;\; \cdot \;\; \cdot \;\; \cdot \;\; \cdot \;\; \cdot \;\; \cdot \\ \qquad \text{there is something which dependently implies 'I shall accept} \\ \hline \longrightarrow T\text{' ([42], p. 210)} \end{array}$$

But as it stands this account is incomplete. For, recalling the discussion of the previous section, the practical argument culminating in the conclusion

> I shall accept T

has two major premises, the proximate and the ultimate. The proximate major here is

> I shall accept frameworks which are Φ.

This ultimate major is

> I shall bring about E.

The sense in which the ultimate major entails the proximate major constitutes one of the most interesting facets of Sellars' theory. Why should I accept the policy contained in the statement of proximate major? According to Sellars, it is analytically implied by the ultimate major of the practical argument implied by the conclusion to the probability argument,

> T is probable$_t$.

The whole purpose behind using probability arguments is to be able to produce explanations; to give explanations one must be in possession of the means to offer explanations, i.e., one must be in an explanatory framework. Now, using probability arguments, one ends up with conclusions whose anaylysis says you can formulate a practical argument incorporating the results of your use of the probability reasoning, with the conclustion,

> I shall accept that-p.

The existence of such an argument is entailed by using probability arguments in the first place, otherwise why would you have sought a conclusion of the type that results from such reasoning? By using probability arguments you already commit yourself to the existence of the practical argument. By committing yourself to the practical argument you commit yourself to the purpose behind the practical reasoning, which constitutes the basis for your having used the probability arguments in the first place. Thus you are committed to some general goal, which goal entails a policy toward accepting the results of probability arguments. If this isn't a circular argument then perhaps there is nothing wrong with Sellars' account. But consider the steps.

1. Accepting a non-metric sense of 'probable' entails accepting the purpose behind probability arguments.

· 2. Accepting the purpose behind probability arguments entails accepting the general policy E.

3. Accepting E entails accepting the subpolicies entailed by E.

4. Accepting both E and its subpolicies entails accepting the use of practical reasoning.

5. Accepting the use of practical reasoning means recognizing that the availability of a probability argument entails there are good reasons to accept the conclusion of probability argument since this is entailed by the meaning of 'probable'.

To place this all in the context of the reasoning surrounding the acceptance of a theory means that we use probability arguments because we wish to arrive at a conclusion concerning the logical relation which obtains between an hypothesis and some data. But we want that information in the first place to help us achieve some goal, E. Thus probabilistic reasoning is goal oriented. The specific goal in mind here is that of being in an explanatory framework.

But why should one accept the policy? By what end is it analytically implied? Surely the state of being in possession of such frameworks logically implies accepting such frameworks *if one does not* already have them. And that this state is the end in question is supported by the fact that it simply unpacks the concept of being able to give non-trivial explanatory accounts of establishes laws. ([42], p. 210)

The purpose of using probability arguments of mode T is to obtain an explanatory framework. This entails accepting F_e if one doesn't already have one. It also entails rejecting the explanatory framework we are using if it is not identical with the one mentioned by the conclusion of the practical argument.

An explanatory framework, F_e^1, tells us what there is. If inaccurate or incomplete then it will at least permit the seeking of a more adequate one, F_e^2. F_e^2, once obtained, will replace F_e^1. Thus, not only can we look for a new F_e, even if we already have one, but once we find it we can replace the old framework with the new.

This leaves us in the position of saying that we are justified in accepting a theory if it succeeds in bringing us closer to our goal of being able to give explanations (à la Sellars). But, according to Sellars, a full explanation requires a final theory, FT. Therefore it would appear as if we are not justified in accepting a theory if the ultimate goal, FT, is unattainable. The case for the goal oriented nature of science appears weakened. We can retain the teleological flavor however even if it turns out that FT is unattainable, and we abandon our search for FT. First, it does not follow that we are unjustified in accepting a theory if *FT* cannot be constructed. Since the justification given above for accepting a theory is that it brings us closer to giving explanations, we can go right on rejecting theories that are inadequate if we can also construct theories that at least tell us more about what there is than old theories have been able to do.

Second, Sellars does not claim that *T* must be true in order for our acceptance of *T* to be justified. If we could show that *T* were true we would be *validating T*. But the validation of theories is a different problem than the justification of a policy concerning the acceptance of theories. The validation of *T* entails conclusively verifying *T*. It has been well argued that we can never completely verify an empirical theory. One simple reason concerns the fact that verification involves showing that all possible evidence for the theory favors the theory. Since the future remains indeterminate, the future evidence is unavailable, hence the body of evidence is incomplete.

But that argument does not rule out the possibility of justifying a *policy* toward the acceptance of theories. Rather than concern himself with the problems of verifying or *validating theories*, Sellars directs his attention to *vindicating policies* to adopt theories constituting more complete explanatory frameworks. If following the policy produces the desired results, we are justified in following the policy. If we do not obtain the desired results then we were not justified using that policy. The ends justify the means.

But even if following a given policy does produce the goal how can we justifiably claim that continued use of that policy will produce continued good results? We can't. Vindication remains a very weak form of justification.

But to look at the problem of justification in this light is to place the burden on the construction of a theory without fully appreciating the role of a theory: explanation. Thus, it would be more appropriate to say that the vindication of inductive policies is the explanation we can offer using the theory the policy helped develop. In this way we put the burden of the justification job on the explanatory function of the theory.

If we now recall that different modes of probability are said to stand in a logical order of dependence, with the probability of theories resting on nomological probability, we can begin to understand the full import of the claim that,

... the end-in-view in *nomological* induction (which must not be confused with the mode of induction involved in the statistical or proportional syllogism to be explained in a moment) is not the possession of empirical truth, but the realizing of a logically necessary condition of being in the very framework of explanation and prediction, i.e.

being able to draw inferences concerning the unknown and explanatory accounts of the known. ([42], p. 219)

If, as a matter of fact, we can make true inferences from the known to the unknown and give explanations, i.e., tell what the thing in question is, then, for Sellars, we are vindicated in our use of the policy which says: accept those theories which are Φ. And despite the unhappiness in Sellars' account of 'probable', this is worth saving.

To develop this point in some more detail let us consider the mode of probability called 'nomological'. To begin with, lawlike statements, for Sellars, are rules of inferences. They are not to be confused with universally quantified material implications or statistical claims. A law is not an empirical claim. The point of nomological induction, the form of induction located in the probability argument leading to the conclusion

h is probable$_n$,

is to arrive at the terminal outcome:

the acceptance of h.

The purpose of nomological induction is the espousal of rules of inference which are lawlike statements.

Now the inductive move does not occur within the probability reasoning itself. Nor can we locate it in the practical argument. It is manifested in the shift from the claim that there is a logical relation between the hypothesis in question and a body of evidence,

$R_L (h, e)$

to the claim

h is probable$_n$.

That this constitutes the inductive move must be the case for two reasons. First, nothing else could count as induction as opposed to probabilistic inference in the strictest readings of those concepts. Sellars correctly cautions from the start concerning the difference between probability reasoning and inductive reasoning. As a matter of fact he goes so far as to claim that it is a mistake to approach induction through the notion of probability, where such a program attempts to reduce induction to nothing more than probability. Secondly, his argument against Carnap's purely logical theory of probability, which cautions against reducing the ground of probability to probability itself, can only be interpreted to mean that the strictly logical relation does not entail

h is probable.

Consider what he has to say about Carnap.

63. What is the relation of my account of Carnap's? A fundamental similarity is that on neither account is there such a thing as an inductive argument of which the *conclusion* is the hypothesis characterized as probable. In this sense there is no 'rule of detachment'. On the other hand we differ in that on my view the assertion of the hypothesis is the 'terminal outcome' of rational sequence of events the occurring of which is the whole point of the inductive enterprise.

64. We agree that the fact that the evidence makes the hypothesis probable is, given that the evidence obtains, a matter of logic rather than of empirical fact – though I would emphasize the presupposition of a major practical premise. On the other hand, his account of inductive argument never reaches the practical reasoning which is its core, but remains at the level of second order probability statements and arguments. From my point of view, he fails to see that

h is probable on e [because R_L (h, e)]

is a second level probability statement of the form

$$\left| \begin{array}{l} \textit{I shall bring about E.} \\ R_L \ (h, e) \end{array} \right.$$
$$e \quad \longrightarrow \text{It is probable that } h$$

the ultimate *inferential* cash value (practical outcome) of which is the reasoning

e
therefore I shall accept h.

Still, the *ultimate cash value* (terminal outcome) is the state of accepting h which results from carying out this reasoned intention. Now given that Carnap overlooked this practical dimension there was, as I see it, nothing to prevent him from making the serious mistake of reducing

h is probable on e [because R_L (h, e)]

to

R_L (h, e)

and thus identifying a *ground* of probability *with probability itself*. By doing so he turns the analytic connection between probability and the intention to accept a hypothesis into a puzzling synthetic connection between a logical fact

R_L (h, e)

and the reasonableness of accepting h. This leads him into attempts to explain the reasonableness of accepting h in terms of the reasonableness of a policy of action, e.g. betting, whereas the truth of the matter, as I see it, is that the reasonableness of policies of action is to be explained in terms of the reasonableness of accepting hypotheses about the outcome of action. ([42], pp. 219–220)

If I am right, the second order probability statement Sellars refers to incorporates the crucial inductive move at issue. The ground of probability remains the logical relation. Probability itself involves the assertion of the reasonableness of accepting h, a function of the logical relation but not the relation itself. Once the move has been made from the logical relation to the claim that

> h is probable,

then the practical dimenstion is implied by virtue of the meaning of 'probable'. In that case, to conclude that h is probable, and hence reasonable to accept, on the basis of a purely logical relation, is to make an inductive inference. From the knowledge that a certain logical relation obtains, we infer that something can be asserted on the basis of an argument implying that knowledge.

But now let us return to the explication of nomological probability. Working backward from the conclusion of the practical argument resulting in

> I shall accept 'LL'

(where 'LL' is the abbreviation for some lawlike statement) Sellars eventually states the goal of nomological induction.

51. Clearly the practical argument for which we are looking culminates in something like

> n/m examined As are B therefore I shall accept 'that K is an unexamined class of As implies that approximately n/m K is B'

and this points to a premise which expresses a general conditional intention, thus roughly,

> I shall accept 'that K is an unexamined finite class of Xs implies that approximately n/m Ks are Y', if n/m of the examined Xs are Y

or, to represent this premise in a way which explicitly *mentions* something which is (partially) *shown* by the above formulation,

> I shall accept a P-implication if it *accords with the observed facts*

Needless to say, this highlights the question 'What is it for a P-implication to *accord with* observed facts?' The answer to this question is implied by the answer to the question 'What is the end-in-view with respect to which this policy is the logically necessary means?'. For on our general account of probability, the practical reasoning for which we are looking must have something like the following structure

>I shall bring about E
>
>But bringing about E logically implies accepting a P-implication if it accords with observed facts
>So I shall accept a P-implication if it accords with observed facts
>'That K is an unobserved finite class of As implies that approximately n/m K is B' accords with the observed facts
>Therefore I shall accept this P-implication

52. What then is E and how does it explain what is meant by saying that a law-like statement *accords with* the observational evidence? The answer is surprisingly simple

>E is the state of being able to draw inferences concerning the composition with respect to a given property Y of unexamined finite samples (ΔK) of a kind, X, in a way which also provides an explanatory account of the composition with respect to Y of the total examined sample, K, of X. ([42], p. 215)

Thus the sense in which the probability of theories depends on nomological probability is the sense in which having rules of inference is a prerequisite for accepting a theory wherein we can reason from the known to the unknown. This also explains the sense in which Sellars and Goodman agree on the principle that it is reasonable to make inferences on the basis of generalizations for which we have no disconfirming evidence. With this in mind we can understand how the policy to accept theories which are Φ can be vindicated. The policy can be vindicated if indeed we can make those inferences and offer those explanations for which having a theory is logically necessary. The policy is vindicated if by following the policy we end up accepting a theory in which sound inductive inferences are possible. This also is the goal of Goodman's theory of projection.

Returning to the different modes of probability, only a few comments need to be added to our present discussion. The question: 'Which rules of inference should we accept?' remains unanswered. The decision to espouse a rule of inference occurs in the context of attempting to realize being in F_e. Hence, we accept only those rules of inference which contribute to that end. We are justified in acting if following that policy does in fact put us in F_e.

As can be easily noted from the quote giving Sellars' reasoning concerning the practical argument appropriate to nomological probability, there is also another dimension to the acceptance of rules of inference. We accept rules not just to be in any F_e, but to be in a framework appropriate to phenomena of a given kind.

Rules of inference mention certain empirical claims. The formulation of those empirical claims remains a function of another mode of probability, the proximate conclusion of which is the claim that a certain empirical regularity is probable in a given degree. This mode of probability is a logical prerequisite to nomological probability insofar as these claims are a part of lawlike statements. But, with respect our interests, Sellars' discussion of this aspect of his theory would carry us far afield. A more important point for our purposes is that a metric analysis of probability supposedly enters at the level of formulating empirical claims. Sellars explains the relation between this metric notion and the basic sense of 'probable' at work up to this point in the following way.

68. The fundamental rationale of this connection is bound up with the fact that although

$$\text{prob}\,(h,\,e) = 1/4$$

doesn't warrant

I shall accept h

it is, on the other hand, true that where

$$\text{prob}\,(h_1,\,e) = 1/4$$
$$\text{prob}\,(h_2,\,e) = 1/2$$

then

$$\text{prob}\,(h_1\,v\,h_2,\,e) = 3/4$$

and this *does* warrant

I shall accept '$h_1\,v\,h_2$'

69. Thus the meaning of the metrical function is logically tied to the basic concept of probability as 'reasonable-to-acceptness'. Over and above the tie between

$$\text{prob}\,(h,\,e) > 1/2$$

and

e makes it probable that-h

and hence between the former and

it is probable that-h

as well as

I shall accept h

the apparatus of statistical and proportional probability is largely the logic and mathematics of combinations, and the chief philosophical interest concerns the ends-in-view and entailed policies which support the apparatus. ([42], p. 221)

The crucial claim here is that the meaning of 'metric function' is tied to the basic notion in such a way that a probability of over 0.5 automatically warrants the conclusion that one shall accept a conclusion which has that probability. But since, as we have seen, Sellars denies that there is a rule of detachment at work in his theory, we must at least try to uncover the secret to understanding what *is* going on there.

An explication of the sense in which the non-metric and the metric senses of 'probable' are tied together is the best we can do. Let us read Sellars' claim that

$$\text{prob } (h, e) > 1/2$$

is required before it follows that

I shall accept that-h

in the following way. We use probability reasoning to achieve a certain end. The attempt to achieve that end is implicit in the appeal to probability theory. As such, the policies which are a logical consequent of pursuing that goal influence our actions in what we might call the explanatory probability-context. The policy to accept a given statement on certain conditions is therefore implicit in arguing that only a hypothesis which stands in a relation to its evidence such that the probability of the hypothesis on the evidence is greater than 0.5 can lead to the conclusion that

I shall accept h.

Thus, we are to read 'probable' strictly. If a given statement is probable, then there exists a practical argument in its logical neighborhood which has as its *terminus ad quem*

I shall accept h.

In reconstructing the practical argument we must pay particular attention to the proximate major. For there we find stated the conditions, C, under which h is to be accepted. Incorporated in C we find the requirement that the logical relation established between the hypothesis and its evidence be greater than fifty percent. We should only accept those empirical claims for which there exists a good chance of reasoning correctly about the remainder of their data, which remainder remains unexamined. We desire to be in a position where we can infer on the basis of what we know to what is as yet unknown.

$$\text{prob } (h, e) > 1/2$$

constitutes the minimal condition under which this objective can be accomplished.

But, that this really does amount to a minimal condition is demonstrable only if accepting the policy does indeed produce F_e. And even then that such a policy helped produce one F_e hardly justifies our belief that it will continue to do so if used in similar endeavors. That is, we have no *rule of detachment* which asserts that if the probability of h on e is greater than 0.5, accept it as true. Instead we have a *policy* that says we should accept hypotheses with that degree of probability because they represent the minimal empirical condition which we can use to produce F_e. Of course we don't know that we can produce F_e by following this policy, but if we can then the policy will be vindicated.

Effectively this amounts to stacking our policies. If we vindicate the policy that ends up putting us in F_e, we vindicate the policies which we used in leading up to that situation. But, once again, that still does not warrant continued use of those policies.

The problem with all this is there are too many unacceptable conditions involved. If you accept the meaning of 'probable' Sellars advocates, you are compelled to use the straight rule even when there is no constructible practical argument for accepting the conclusions of first order probability arguments. But, we were told that taking the basic meaning of 'probable' entails there are such practical arguments. Thus we are left in the position of accepting that probability arguments entail practical arguments for accepting conclusions of probability arguments even when no such practical argument can be constructed, i.e., in the case of first order probability arguments which are incomplete in principle. Now we might be able to live with this by inventing an account of first order probability arguments which places them in the realm of something like a subliminal thought pattern or some such and manage a fudge however inelegant. But the issue is more complicated than that. For Sellars requires the rationality of the inductive mode of inference. The vindication of this presupposition is the completion of a final theory whose explanatory capacity will justify the initial acceptance of the general policy of postulating unobservables to explain observables. This is merely another conditional argument. And the continued 'iffy' nature of the explanationist enterprise raises serious doubts about the viability of the general program.

But these doubts would only be justified if the explanationist elements in Sellars' analysis were responsible for these results. For as we have noted above there are two dimensions to Sellars' theory of justification. First, there is the vindication of the inductive policies. But, again, vindication is a weak method of justification. It is supplemented by the explanationist principle of justification: a theory is justifiable if it provides explanations.

To fully appreciate these different stands, and thereby attempt to salvage some of the ideas, requires recalling that much of what Sellars has to say on these topics is influenced by his efforts to separate the process of scientific inquiry from the product of science. Sellars' teleological view of the development of science leads him to *vindicate* policies used in the process of developing theories. Science is concerned with developing theories postulating unobservables. We vindicate our techniques for constructing theories by producing theories. But given the theory, what justifies our assumption that what it claims ought to be believed? The criterion at work here, again a function of the teleological view Sellars has of science, is that we justify our adoption of a theory, and hence our belief in what it says by showing that the theory explains what we would otherwise believe.

Explanationism claims that a theory is justified if explained or if it explains. As it stands this not only sounds reasonable, but worth developing. Explanation is epistemic in nature and the appeal of explanationism lies in the doors it opens. We have been stymied in our attempts to justify theories by appeal first to syntactic and semantic principles of verification and then confirmation. Now, by enlarging our context we may be able to finally break some new ground.

On the other hand, it is not clear that distinguishing between policies (process) and theories (product) really does leave us in any better position. These doubts arise because the distinction between justifying theories and vindicating policies is not as firm as we would have it once we reject the idea of an FT. For we do not vindicate policies which we have used to construct just any theory; rather, we only vindicate policies for a theory which can function as an explanatory framework. Therefore, if a theory is adopted at time t_1 and rejected at time t_2 because it has been replaced by a superior F_e do we thereby rescind our vindication of our inductive policies? Sellars' safeguard against this situation was his suggestion that the final sense of 'vindicate' applies legitimately only in the case where we produce FT. Without FT we are forced to talk about the vindication of policies with respect to particular theories, and the claim of extreme contextualism again seems reasonable. More about this in Chapter V. Now it is time to turn to an analysis of 'theory'.

CHAPTER IV

THEORIES

1. INTRODUCTION

As we have seen, Sellars' views on induction and probability are directed toward the construction of a theory adequate to explaining the world around us. I would now like to consider specifically Sellars' account of theories by looking as their structure and their role. What Sellars has to say about each of these topics bears on what he has to say about the other, hence the need to consider both.

Sellars' own account of the structure of a theory emerges from his attack on the Positivist analysis. As usual he has many axes to grind. One can often lose sight of the end by failing to realize that, for example, Sellars not only is attacking Nagel on the role of predicate variables, but he is also developing a theory of meaning which in turn is not only a response to the positivists but the foundation for a general theory of empirical content. Finally, it must be remembered that Sellars is constructing a framework in which the development of science can be characterized and that it is his emphasis on the activity side of the coin that often places him at odds with others writers.

Sellars' two major objections to the Humean analysis of the structure of a theory are that it is incomplete and epistemologically unsound. These two points are independent, i.e., the epistemological failure cannot be simply corrected by filling in the missing elements of the description.

The charge of incompleteness results from the Positivist preoccupation with the product of scientific inquiry. According to Sellars, science does not merely strive to produce theories, it develops theories for people to use. The formal descriptive structure to which Positivists limit their attention requires an addendum to those formation rules which tell us both how to add new terms to the theory and what constitutes a well-formed claim within the theory. We also require rules which specify under what conditions using the theory proves profitable, i.e., when we can expect an explanation to be forthcoming. These rules, like formation rules are *about* a given structural aspect of the theory, not part of that structure. They meta-level statements. According to Sellars, a theory involves both an object-level and a meta-level.

He completes the Positivist's account not by distinguishing these two levels, but by locating such items as formation rules, which Humeans blindly class together with other components of a theory, at the meta-level.

But Sellars is not merely concerned to reestablish the principle that theories are to be used. The major point here turns on the explanationist principle that an adequate scheme of *justification* entails such sensitivity to the use of theories. An adequate account of a theory must provide for both the role of a theory and its method of justification. The relation between these points is such that it necessarily affects the description of a theory. The role of theories is explanation. The rules of the meta-level tell us what we have to do if we are to produce explanations. And only if the theory can produce explanations can we say it is justified.

The charge of epistemological inadequacy stems from two sources: Sellars' Kantian bias and his program for the development of science. The Kantian influence is manifest in Sellars' exchange with Nagel over the need for predicate constants within the theoretical vocabulary of a theory. Sellars' argument amounts to a partial rerun of Kant's dictum that percepts without concepts are blind. But this is a difficult point to sort out. It looks, at first, as if he claims that on Nagel's view a theory cannot be empirically tested. But on a closer analysis it turns out that Sellars' real point is not that theories cannot be tested if Nagel is correct, but rather that unless theories contain predicate constants they are eliminable. They are eliminable because they can be reduced to the empirical level. And, if eliminable, no fully scientific image of man can develop out of the manifest. The only way to develop an improved image of the world is if postulated theoretical entities are assumed to be real. This way they cannot be reduced to the objects of the manifest image. Predicate constants play a key role in insuring the irreducibility of a theoretical picture to that generated by the framework of the manifest image.

Clearly then predicate constants have a role to play in Sellars' program for a developing science. Moreover, Sellars has two roles in mind for predicate constants. The first has already been discussed, viz., predicate constants provide a guarantee against the irreducibility of theories to the observation framework.

Secondly, appeal to predicate constants provides a convenient structural starting point for a theory of the development of science. According to Sellars, science develops piecemeal by replacing discredited claims. ([45], p. 21, and [46], p. 360.) Correspondence rules provide the means by which this can be accomplished. They are used to redefine old concepts in terms of new ones. The essential concepts involved in this redefining process are characterized by predicate constants. They play a special categorical role in the development of science. Sellars appears to believe that certain concepts have a lien on truth. They mark out crucial areas for inquiry. Science progresses by the continuingly more specific articulation of those concepts within a coherent system. This is insured by securing ever more adequate auxiliary concepts to help clarify the key categorical ones.

How to determine which of the many theoretical concepts of science are these key concepts of Sellars' is a very difficult question. The answer lies somewhere in the context of his discussion of their role. He belives that the key concepts in science can have their history traced. " ... scientific terms have as a part of their logic a 'line of retreat' as well as a 'plan of advance' ... " ([40], p. 288) Once we develop a final science those key concepts in that theory will mark the culmination of the development of their intuitively valid but incomplete or overly wrought ancestors. Thus, it appears that we will not know which concepts are the really important ones until we finish building a completed science. This is an unfortunate drawback. But it leads to the following interesting possibility. If Sellars is correct concerning the necessity for predicate constants, then it would appear possible to develop a method for analyzing the history of science which takes its clue from the idea that the development of science can best be understood as successive attempts to work out the significance, and maybe even the 'logic', of a set of specific notions. This program would entail a detailed case by case study of science.[1]

In his discussion of theories, Sellars pays a good deal of attention to the standard contemporary Positivist interpretation of the issues at hand. He acknowledges the significance of this analysis and its value in illuminating problems. He balks, however, at accepting the conclusion of the argument based on this analysis. To begin with, he does not admit a neutral observation language, NL_o. Such neutrality follows if the uninterpreted theoretical vocabulary of a theory obtains its significance only by being tied to an independent observation context.

Having located the objectionable consequence, Sellars suggests the reason for the unhappy result lies in the improper interpretation of one of the premises in his opponent's argument. This is a standard Sellarsian move. (A classic example of this type of dialectical presentation is [47].) He usually then isolates the factor at fault, proceeds with a few modifications here, some additions of his own there and, while apparently still operating within the context of whatever position he started out opposing, draws a conclusion which if not straighforwardly contradictory in terms of that position, is at least incoherent in that context. Only in reexamining Sellars' own argument does the illusory nature of the impression that Sellars ends up within the context of the position he started our examining become clear. While making those few modifications he shifts from arguing the merits or faults of some alternative analysis to producing his own interpretation of the nature of the problem and a solution.

This dialectical method of analysis, although not always apparent, permeates Sellars' discussion of a theory. Following DN, he divides a theory into three parts. ([44], pp. 106–107) On each of these components several methodological changes are imposed. These changes provide the grounds for rejecting the Positivist's method of analysis. He ends up describing theory in terms not shared by DN. For example, he eliminates NL_O. The threefold division of a theory remains after Sellars has finished his analysis, hence the illusion that we are still talking about DN. But the significance of each of these components as well as their structure has been altered. The set of consequences which follow from speaking about a theory in this fashion differs critically from that which follows from DN.

2. THE SELLARSIAN VIEW OF A THEORY; AN INTRODUCTION

The type of theory with which Sellars deals "postulates unobserved entities to explain observable phenomena". ([44], p. 106) He assumes "that *something* like the standard modern account of this type of theory is correct". ([44], p. 106) Continuing along this line he distinguishes between:

(a) the vocabulary, postulates, and theorems of the theory as an uninterpreted calculus;

(b) the vocabulary and inductively testable statements of the observation framework;

(c) the 'correspondence rules' which correlate, in a way which shows certain analogies to translations, statements in the theoretical vocabulary with statements in the language of observation. ([44], p. 107)

But, for Sellars, a theory involves more than just an interpreted formal system. Theories are the means by which we come to know what there is. In Sellars' account, the complete structure of a theory comes closer to that of a game than a simple deductive system. And as in a game, it is not the case that mere use of deductive logic will get you where you want to go. Pragmatic rules of application must, therefore, be added to the above account. Moves in a game are determined by the objective of the game; methodological rules of strategy tell you how to achieve the objective.

To put the matter as simply as possible, a theory consists of a formal system embedded in a more complex system of practical reasoning concerned with the *use* of the formal system for the particular purposes of explanation. We can also describe a physical theory as a complex language consisting of an object language concerned with causal relations and a meta-language couched in terms of ought-statements. The latter provide directions on how to use the oject language for explanation, description, and prediction.

A theory cannot be merely a formal system. When we reject NL_O the need to place theories safely in the hands of the people who design and use them becomes crucial. For what sense does it make to speak of a formal system which cannot be shown to be better or worse than any competing system and whose relation to the world is at best minimal? Locating theories in the context in which they are used remains the only move left. Theories are tools made by and for the use of people. How people use and justify these tools is critical to their proper description. We must not only reintroduce the notion of the practical reasoning surrounding theories, but also reconsider the nature of the goal behind the use of theories.

Theories are constructed to help us determine what there is. This constitutes the basis for a number of Sellars' attacks on the Positivists. That tradition approaches the question of the nature and structure of theories in terms of how to consider their product, knowledge. Sellars claims that this is not far-reaching enough. We assume that science provides us with knowledge and guarantees must be made for that point in our analysis. An inadequate analysis precludes the possibility of science providing knowledge. The Positivist analysis without the use of neutral observation language precludes empirical knowledge. But important as it is in the long run this is just a side issue. The crucial point emerges in the guise of securing the success of science.

This success will be marked by producing a complete scientific theory. The criterion of completeness here is explanatory coherence. How this goal can be reached and the justification for actions designed to accomplish this end constitutes the major Sellarsian problem and some of his views here were discussed in Chapter III.

3. SELLARS AND NAGEL ON THE FORMAL STRUCTURE OF THEORIES

It seems appropriate to begin with the notion of a calculus, the core of a theory. Sellars attacks Nagel on this point. ([46], pp. 341–358) The conflict between Nagel and Sellars centers on whether or not the theoretical vocabulary of a théory contains predicate variables and predicate constants. Nagel opts for the predicate-variables-only analysis. He believes that,

... a fully articulated scientific theory has embedded in it an abstract calculus that constitutes the skeletal structure of the theory. ([24], p. 91).

By viewing the nonlogical terms occurring in key statements as constituting only place holders and by concentrating on the structure of the statement and logical relations involved, Nagel believes the calculus can be abstracted from the theory.

The nonlogical terms of a theory can always be disassociated from the concepts and images that normally accompany them by ignoring the latter, so that attention is directed exclusively to the logical relations in which the terms stand to one another. When this is done, and when a theory is carefully codified so that it acquires the form of a deductive system ... the fundamental assumptions of a theory formulate nothing but an abstract relational structure. ([24], p. 91)

It is one thing to assert here that the fundamental assumptions of a theory can be viewed as an abstract relational structure, but quite another to go on to say that:

the fundamental assumptions of a theory constitute a set of abstract or uninterpreted postulates, whose constituent nonlogical terms have no meaning other than those accruing to them by virtue of their place in the postulates, so that the basic terms of the theory are 'implicitly defined' by the postulates of the theory. ([24], p. 91)

That a theory can be analyzed into abstract relational statement-forms (as Nagel goes on to call them) does not thereby license the claim that the fundamental assumptions of the theory are nothing more than these logical structures.

The second point is slightly more difficult to isolate. Given Nagel's claim that the nonlogical predicates of a theory only obtain meaning by implicit definition, it *appears* as if Sellars' objection comes to the question of how theories can be tested. Moreover, it further appears as if Nagel has worked himself into a corner from which neither a theory of verification nor a theory of confirmation can extradite him. For, as the situation seems to develop, the challenge does not concern a theory of evidence, but rather the general issue of how to show that the type of theory Nagel has described here is capable of even being subjected to empirical confimation. The theory is an abstract calculus whose nonlogical predicates are implicitly defined in terms of the postulates which are uninterpreted. If the postulates are uninterpreted, how do we tell what constitutes evidence for them? Or, how can you have testable instances of statement forms? If, however, the predicate expressions of a deductive system are essentially variables, the statement functions of the system could have no statement counter-part with predicate *constants* for which they were true – surely an incoherent notion. ([46], p. 341.)

If we transfer this line of thought to theories about objects in the world, the point becomes: unless there are predicate constants in the postulates which form the basic assumptions of the theory then there could be no possible tie between the theory and the world.

In order for there to be a tie between the theory and the world the postulates of the theory must contain predicate constants because they are the basis for evidential relevance. They serve to signal the observer about the relevance of certain data to the theory and its possible role as an evidential instance, in this way providing the grounds for relevance in terms of empirical significance, as opposed to the meaning of the constant.

On this proposed account, the meaning of the constant is determined in a two-fold manner. First, the constant is implicitly defined by the structure of the calculus and the subsequent relationships into which it can be shown to stand with other predicates. In this sense Sellars would appear to agree with Nagel on the nature of implicit definition for predicates of the non-logical vocabulary of the theory.

But Sellars goes on to require that some of the predicates be constants. Predicate constants are already meaningful concepts and, hence, not open to indiscriminate substitution. They have established roles to play within the theory. The types of inferences possible on the basis of the constants restrict the generality of the calculus since, given that they are meaningful constants, certain relational patterns with other constants are antecedently established. This restricts possible substitutions to variables.

On the point of substitution for variables, consider the analogy between theory constants and category words. The names of two categories cannot be interchanged in a category scheme and the remainder of the category remains as it was because what can be included under one category may be either irrelevant or meaningless within another. The according of improper data to a category is a Rylean category mistake, as in asking for the university after you have seen the buildings, met the faculty, staff and students, visited the library, seen the campus, etc. Again if the two categories, say space and substance, were to be switched within a physical theory, space concepts now would fall under what was formally the concept of substance; the iteration of the new data with that leftover from the old ordering could create conceptual chaos.

Insofar as predicate constants function analogously to category words they cannot be interchanged. Let us be clear about the manner in which predicate constants affect the theoretical framework of a theory. For Nagel,

insofar as the basic theoretical terms are only implicitly defined by the postulates of the theory, *the postulates assert nothing*, since they are statement-forms rather than statements (that is, they are expressions having the form of statements without being statements), and can be explored only with the view to deriving from them other statement-forms in conformity with the rules of logical deduction. ([24], p. 91, italics mine)

Postulates are merely structural devices for allowing certain logical manipulations. For Sellars, on the other hand, postulates not only establish certain logical possibilities, but they also are meaningful statements. And if provided with an empirical interpretation they can be claims about the world. We exclude some moves within the theory because of the relations established by using predicate constants. These limitations would not occur if indescriminate substitution of variables were possible. The cash value of this approach is, for example, the elimination of the aether once the concept of gasses was fully articulated. But, given an interpretation, Nagel's postulates can also be said to be about the world.

The specification of the relations due to predicate constants plays an important role in Sellars' long range view. The development and progress of science involves replacing concepts by their more fully developed counterparts (what Sellars calls successor-concepts in [48]). If this procedure works, then it would be possible to trace the development of a given concept as the theories become more sophisticated. This is an interesting notion and goes a long way towards explaining Sellars' insistence on predicate constants. Laid bare the assumption states that some concepts form a more fundamental grouping than others. The group contains those concepts which when fully developed, as a coherent family, form the basis for a successful theory.

The second way in which the predicate constants obtain meaning involves their connection with extra-linguistic objects. Several interesting points arise in connection with this facet of Sellars' analysis. On the one hand, he uses this account to drive the final nail into the coffin of predicate-variable-only. On the other, it leads to a consideration of observation languages and correspondence rules. First, however, let us examine the connection between predicate constants and the world.

Sellars concludes his discussion of the need for predicate constants in the uninterpreted calculus by considering the following case: for the predicate expressions of a pure geometry, we must "specify a primary use in which they constitute a special class of predicate constants". ([46], p. 342) The role these predicates play constitutes the grounds for determining membership in this class.

Their role is not that of being substituted for or quantified over, but that of being available for connection with extra-linguistic fact. ([46], p. 342)

A pure geometry becomes a physical geometry by connecting the formal predicates of the geometry with empirical predicates. The empirical predicates relate to directly observable aspects of physical objects. "... a nonlogical predicate constant which isn't connected with extra-linguistic objects is not, in the full sense, meaningful. But not being extra-linguistically meaningful must not be identified with being a variable". ([46], pp. 341–342)

Thus, there are two aspects to the meaningfulness of a theoretical predicate: the use of implicit definition, as in the case of predicates in pure geometries and tying these predicates to empirical predicates by some device such as correspondence rules to provide them with a means of exhibiting empirical significance. But, according to Sellars, for a predicate not to be extra-linguistically meaningful, i.e., not related to some empirical predicate, is not sufficient to class that predicate as a variable. Nagel holds that it is. Sellars objects because he believes that to be meaningful is not merely to be empirically significant. A predicate can also be meaningful if it has a well-determined role in a system.

Sellars suggests that non-empirically-significant predicates be considered *candidate constants*. If we can use some appropriate set of rules to transform a non-empirical system into an empirical system, then those predicates which were previously only defined implicitly with empirical consequences would become empirically significant. They are, therefore, candidates for empirical significance.

As I see it, then, just as the predicate expressions of a pure geometry construed as a system of implicit definitions, are not variables but candidate predicate constants, so the predicate expressions in a microtheory *qua* deductive system are not to be construed as variables, but as *candidate predicate constants*, which, to put the matter in first approximation, get their extra-linguistic meaningfulness from the correspondence rules which connect them with constructs in what Bergmann has called the 'empirical hierarchy'. ([46], p. 342.)

Locating the differences between a variable and a candidate constant is difficult. Neither are empirically meaningful, though both can become empirically significant by using some set of rules to connect them with empirical concepts. One difference, however, lies in the nature of the rules by which variables are instantiated and predicates interpreted. Nagel wants a deductive connection between the empirical and non-empirical predicates. The correspondence rules show how to define the non-empirical in terms of the empirical. For Sellars, correspondence rules redefine the objects of the observation framework in theoretical terms. ([44], p. 125) However, the difference in the interpretation of 'correspondence rule' does not crystallize the need for distinguishing candidate constants from variables.

To resolve the issue here let me suggest first that the terminology is misleading and, secondly, that the issue is not one of mere empirical significance. The predicates in question are not candidate constants, i.e., candidates for constanthood. They are *constants* which are candidates for extra-linguistic meaningfulness, which is not to say they are meaningless. They already operate within a system. They can, however, name objects in the observable domain if some set of rules can be found to correlate them with constants in an already empirically significant theory.

With respect to the second point, that the issue is not one of empirical significance alone, Sellars has two considerations in mind. First, not all meaningful systems must be empirically significant. A system can have meaningful components without empirical content. The *roles* they play within the system determine the meaning of the predicate constants. Other concepts are, thereby, definable and we can construct a complete theoretical system, e.g., a pure geometry.

Secondly, Sellars' characterization of the vocabulary of a theory as containing predicate constants constitutes an important aspect of his general commitment to scientific realism. Nagel, by insisting on the occurrence of only predicate variables in the vocabulary, allows for the Ramsification of a theory and hence its reduction to some observational level. Nagel turns to Ramsification in an effort to counter the objection that theories are mere statement forms. By turning the theory into one existentially quantified sentence Nagel believes it thereby becomes a statement because it makes an existence claim. ([24], pp. 141—142) But Sellars even denies that a Ramsey sentence makes an existence claim.

Ramsification turns a theory into one long sentence in which theoretical constants become quantified variables. ([34]) In this manner all the predicates of the theory become members of the observational vocabulary of the theory. Consider, for simplicity's sake, Hempel's brief characterization of a Ramsey sentence. ([16]) He first sets up an example in which the theoretical vocabulary of a theory T, contains the term 'white phosphorus', abbreviated

'*P*'. *T* also includes an interpretation which specifies several necessary but not sufficient conditions for the observational application of '*P*'. Each of these necessary conditions is independent of one another.

Let those necessary conditions be the following: White phosphorus has a garlic-like odor; it is soluble in turpentine, in vegetable oils, and in ether; it produces skin burns. In symbolic notation:

(9.6) $(x)(Px \supset Gx)$
(9.7) $(x)(Px \supset Tx)$
(9.8) $(x)(Px \supset Vx)$
(9.9) $(x)(Px \supset Ex)$
(9.10) $(x)(Px \supset Sx)$ ([16], p. 214)

We need three other points before we can continue to Hempel's example of a Ramsey sentence. First, all the other predicates are members of the observational vocabulary of *T*. Second, there is one final predicate to consider,

... namely 'has an ignition temperature of 30°C', or '*I*' for short; and let there be exactly one interpretative sentence for '*I*', to the effect that if an object has the property *I* then it will burst into flame if surrounded by air in which a thermometer shows a reading above 30°C .... The interpretative sentence for '*I*', then is

(9.11) $(x)(Px \supset Fx)$ ([16], pp. 214–215)

Third, there is one postulate in *T*:

(9.12) $(x)(Px \supset Ix)$ ([16], p. 215)

With this equipment in hand Hempel proceeds to construct the Ramsey-sentence for *T*.

... the method amounts to treating all theoretical terms as existentially quantified variables, so that all the extralogical constants that occur in Ramsey's manner of formulating a theory belong to the observational vocabulary. Thus, the interpreted theory determined by the formulas (9.6)–(9.12) would be expressed by the following sentence ...

(9.13) $(\exists\varphi)(\exists\psi)(x)[\varphi x \supset (Gx \cdot Tx \cdot Vx \cdot Ex \cdot Sx) \cdot (\psi x \supset Ex) \cdot (\varphi x \supset \psi x)]$

This sentence is equivalent to the expression obtained by conjoining the sentences (9.6)–(9.12), replacing '*P*' and '*I*' throughout by the variables 'φ' and 'ψ' respectively, and prefixing existential quantifiers with regard to the latter. Thus, (9.13) asserts that there are two properties, φ and ψ, *otherwise unspecified*, such that any object with the property φ also has the observable properties *G, T, V, E, S*; any object with the property ψ also has the observable property *E*; and any object with the property φ also has the property ψ. ([16], pp. 215–216)

The result, therefore, of substituting a variable φ, for the theoretical predicate 'P', is the elimination of 'P' in favor of its observational criterion. The theoretical property is eliminated and replaced by a place-holder cashed out strictly in terms of observational considerations. If that is indeed possible, then the claim of the theory concerning the imperceptible entities it postulates is reduced to the observational criteria of the manifest image. Theories, therefore, are eliminable since the postualtion of imperceptibles serves no good purpose. These new entities are eliminated as fast as they are conceived.

Now clearly, if the postulating of imperceptible theoretical entities is to explain perceptible entities this kind of reduction cannot take place. To secure this, Sellars launches a direct attack.

Even though the Ramsey sentence does imply — given that we are willing to talk about the concepts expressed by theoretical terms — the existence of concepts or properties satisfying the conditions expressed by the postulates of the theory, the question whether these concepts are *theoretical* or *observational* is simply the question whether constants substituted for the variables quantified in the Ramsey sentence can be construed, *salva veritate*, as belonging to the observational vocabulary. ([44], p. 117, [46], p. 343)

And, we are led to believe, the answer is 'no'. For the predicate constant which we would substitute for the quantified variable would not itself function in the observational vocabulary directly. It would do so by means of its analysis, i.e., the Ramsey sentence. But that itself marks it off as special. In other words, the Ramsey sentence, properly viewed, helps establish the apartness of theoretical constants; it does not succeed in reducing them to observational terms. This 'apartness' is significant to the extent that it distinguishes the observational from the theoretical domain. In this sense the issue might appear to be merely a question of ontology since one result of this separation is the continued independence of theoretical entities.

The key issue in Sellars' distinction between constant-candidates and variables does not, however, reduce simply to the problem of ontological commitment. Even though one of Sellars' goals has ontological implications, it is secondary at this point. The major objective, hopefully clarified by distinguishing between the meaning of a concept obtained by implicit definition and its empirical significance, concerns the remnants of the verification theory of meaning in *DN*. Non-empirical systems can be meaningful. Moreover there are two classes of non-empirical theories: those which are potentially empirically significant and those which are not. The difference between the two is apparently only a matter of practical difficulties in finding an appropriate set of correspondence rules. The role it plays in a system, and not its empirical consequences, essentially determines the meaning of a concept.

But if Sellars' only worry concerns the verificationist elements left in *DN* he wrongly attacks Nagel. Nagel admits that predicates have to be implicitly defined within a theory. ([24], pp. 91, 95, 106) He, like Sellars, also argues that empirical significance requires that some tie be established between the system and the world, usually by means of descriptions framed in some observation framework. ([24], p. 93, pp. 106–152) In this respect no difference exists between candidate constants and variables. Both are intra-system meaningful, but not empirically significant without being connected to the world by a set of correspondence rules.

4. THE OBSERVATION FRAMEWORK

An observation framework, F_o, incorporates both the level of empirical generalization and a vocabulary of observation terms, L_o. Hempel marks out the terrain of F_o in the following fashion.

1. ... it will be helpful to refer to the familiar rough distinction between two levels of scientific systematization: the level of *empirical generalization* and the level of *theory formulation*.
2. In accordance with the distinction made here, we will assume that the (extralogical) vocabulary of empirical science, or of any of its branches, is divided into two classes: *observation terms* and *theoretical terms*. In regard to an observation term it is possible, under suitable circumstances, to decide by means of direct observation whether the term does or does not apply to a given situation.
3. ... accordingly, the observational data whose correct prediction is the hallmark of a successful theory are at least thought of as couched in terms whose applicability in a given situation different individuals can ascertain with high agreement, by means of direct observation.
4. Theoretical terms, on the other hand, usually purport to refer to not directly observable entities and their characteristics. ([16], pp. 178–179)

F_o is the *context* in which both the formulation of singular descriptions and empirical generalizations takes place. The nonlogical vocabulary used in these claims is composed of terms in L_o. Sellars suggests that there is an equivocation here in the use of 'observational' and 'non-theoretical'. ([44], p. 107ff) One of the main points Sellars makes in his discussion of the levels theory of theories is that there is no absolute level. As we read Sellars here, he argues that it is the commitment to this absolute level which allows Positivists to draw a sharp distinction between 'observable' and 'theoretical', and, subsequently, to argue that 'non-theoretical' should be equated with 'observable'. Sellars seems to be suggesting, moreover, that the assumption of an *absolute* level encourages drawing these sharp distinctions. If so, this is surely a mistake since a non-theoretical term need not necessarily be an observation term. Correspondence rules establish a meaning relation between the theoretical terms of T and components of L_O. If 'observational' and 'non-theoretical' are used interchangeably, and if non-theoretical does not entail observational, then the members of NL_O need not be restricted only to observation terms, i.e., terms which denote observable objects. Hence correspondence rules need not only relate theoretical terms and observational terms. The non-theoretical vocabulary of some theory, T_1, could be the vocabulary of another theory, T_2. If F_O contains non-theoretical, as opposed to only observational terms, then we are not restricted to some absolutely rock bottom observation language NL_O. For the vocabulary of F_O may now contain terms from older theories whose meanings are well-established but which are no longer theoretical, strictly speaking. Also by specifying only that the vocabulary of F_O be non-theoretical, in the sense of explainable by T_1, it is now possible for another well-understood theory, T_2, to operate as an F_O.

T_2, however, must be of a lower level than T_1; this much is born out by the quotes from Hempel given above. Given theories which postulate unobservables to explain observables, we can view T_1 as explaining T_2. If we don't stipulate that the non-theoretical framework have a particular vocabulary, i.e., terms which denote only objects of the visible world, then it would seem that any framework capable of being explained by some such T_1 can be called an F_O.

Up to this point there are few if any problems. But the crunch comes when Sellars considers some of the results which follow from postulating levels of theories. "This calls up a picture of levels of theory and suggests that there is a level which can be called non-theoretical in an absolute sense". ([44], p. 107) Sellars calls the language which would be appropriate to this level the physical-thing language and argues against the view that employs it. His major ploy is the correspondence rule.

5. CORRESPONDENCE RULES (C-RULES)

There must be some way of connecting the terms and sentences in the theory with those of the language which describes the world in observation terms. In this fashion the theory can be seen to be about the world and hence empirically clear. Nagel describes the situation in the following way:

If the theory is to be used as an instrument of explanation and prediction, it must somehow be linked with observable materials.

The indispensibility of such linkage has been repeatedly stressed in recent literature, and a variety of labels have been coined for them: coordinating definitions, operational definitions, semantical rules, correspondence rules, epistemic correlations, and rules of interpretation. ([24], p. 93)

C-rules play essentially the same role for Sellars. They link the theory with statements made in F_o. They do not, however, provide a means for adding empirical content to the theory by, for example, partially defining theoretical terms. As noted earlier, by insisting that some of the predicates in the theoretical vocabulary of a theory be constants, Sellars has already partially solved the problem of content. Content is determined first as a function of the picturing ability of a theory. Second, pictures of the theoretical domain are tied to pictures of the observational or manifest image. Thus Sellars retains the notion of C-rules but, in typical fashion, turns them into a new and different creature. "Correspondence rules in this broader sense are rules correlating theoretical predicates with empirical predicates". ([49], p. 329) Within the general category of C-rules, Sellars distinguishes two types, substantive C-rules and methodological C-rules.

In the case of methodological C-rules, the predicates they correlate with the theoretical predicates "need not pertain to the domain of objects for which the theory is a theory". ([49], p. 330) From this two points follow. (a) While the empirical predicates in question need not be concerned with the objects for which the theory functions, they must be empirical predicates. Thus they must be either "observational predicates or related to them by operational definition". ([49], p. 330) (b) Methodological C-rules do not require a fully developed and articulated theory. In this way Sellars secures their role in the development of science.

Correspondence rules of the methodological type can be built on a very schematic and promissory-notish conception of *how* observable phenomena are connected with the theoretical states. ([49], p. 331)

Sellars used the case of the following C-rule as an example:

(A) Spectroscope appropriately related   ↔   Atoms in region R are in such and
       to gas shows such and such lines   such a state of excitation.

(where the arrow indicates the correlation) the spectroscope is not itself a gas, hence not in the domain of the theory. Moreover, we need not have a complete theory of spectroscopy to be able to use the spectroscope in this fashion.

On the other hand, the spectroscope must be related to a gas in an appropriate manner, and we can certainly say that the property of causing certain spectral lines, which *is* a property, albeit relational, of an object belonging to the domain of the theory, is correlated by the correspondence rule with the state of excitation of the atoms. ([49], pp. 330–331)

Now the sense in which A is a methodological rule should be clear. It establishes a relation between the theory and an instrument which concerns the objects of the theory. It pertains to something which, while itself not 'in' the theory, is nevertheless crucial for the method of testing the theory, or instansing it, or implementing it. In technical problems of confirmation this type of C-rule plays a key role.

The second type of C-rule, substantive rules,

correlate predicates in the theory with empirical predicates pertaining to the objects of the domain for which the theory is a theory, where the important thing about these empirical predicates is that they occur in empirical laws pertaining to these objects. ([49], pp. 329–330)

By virtue of the fact that the empirical predicates must occur in empirical laws which are concerned with the objects which the theory is about, it does not follow that they must be observation predicates, or directly relatable to observation predicates. The distinction between empirical and observational predicates is not, however, as awkward as it might look from this account.

Observation predicates are those which occur in observation reports, for example, 'red'. Empirical predicates pertain to the objects of the domain of the theory. They need not occur in, sometimes it would even be nonsense to suggest that they occur in, observation reports. The key point rests on theoretical empirical predicates occurring in laws, while observational predicates need not in order to continue being considered as observation predicates. Sellars provides the following as an example of a substantive C-rule:

(B) Temperature of gas in region R is such and such	↔	Mean kinetic energy of molecules in region R is such and such

(B), unlike (A), does not provide a path whereby we may go from a theoretical description to an empirical description. It does not incorporate a method for going from one kind of description to another. Rather, it correlates the two sides of the C-rule. The import of this sort of correlation is that,

the rule . . . (B) . . . in some sense *identifies* temperature with the mean kinetic energy of its molecular energy of its molecular constituents, (while) it would be absurd to say of . . . (A) . . . that it *identifies spectral* lines with the state of excitation of the atoms. ([49], p. 330)

Sellars proposes a peculiar form of identity using substantive C-rules. Given B as a correlation of predicates in a theory with empirical predicates, the question arises as to how many types of entities there are in the world. Are there, among other things, both gases and molecules, or only one or the other?

According to the view I am proposing correspondence rules would appear in the material mode as statements to the effect that the objects of the observational framework *do not really exist – there are no such things.* ([44], p. 126)

Correlating statements in this way, a statement results to the effect that the objects to which the empirical predicates pertain do not exist. This, however, leads to the strange situation of denying an object's existence, on the one hand, while asserting it on the other. In what sense is this an identity?

Sellars characterized the identity as one of *use*. The problem generated by characterizing correspondence rules as permitting the claim that a given class of items do not really exist can be solved as follows. The C-rule merely establishes a correlation between the two types of statements. However, by invoking another set of rules, explaining the correlation, the conclusion can be drawn that the objects in the empirical domain do not exist. They do not exist because they are merely the objects described by the theoretical predicates. Here Sellars' views on explanation play a very important role.

Given the C-rule and its role correlating statements in different domains, the basis for the material mode claim that the objects of F_o do not exist lies in appealing to the greater explanatory power of the theoretical framework F_t. F_t is used to explain statements in F_o. But the reverse cannot take place. Thus, the world contains not two types of entitites, gases and collections of molecules, but only one, because gases *are* collections of molecules. The consequence of viewing C-rules in this way, when conjoined to the idea that theories are developed to explain perceptible objects, is that explanation involves an element of essentialism. To explain amounts to describing what really is the case, the explanation of X is: X doesn't really exist, what really is the case is Y. This amounts to a rejection of DN, where explanation, it is claimed, involves deduction.

Sellars' position, reconstructed, comes to this: If we are committed to deduction as the logic of explanation, then we would be committed to the existence of the entities of both the derived and the deriving scheme since we can formulate deductive explanations for each type. But since explaining an entity entails saying what it is, i.e., describing its essence, we have only one entity we are talking about, even though there are several ways to discuss it. Hence the relation cannot be that of deducibility.

> *SE Protothesis*:   Explanation involves both the replacing of a theory, T_1,
> by another theory, T_2, using C-rules, and the denial of
> the reality of the objects of T_1.
>   1. To explain X in T_1 requires:
>   (a) providing the meaning of X in T_2, and
>   (b) a rule to the effect that the entitites X denotes in
> T_1 do not exist.
>   2. To explain the meaning of X in T_1 via a substantive
> C-rule, correlate X with a statement in T_2, e.g., Y.

Among other problems stemming from this account of explanation, the following bears on our discussion of C-rules. If C-rules establish an identity between statements in T_1 and T_2, and explanation essentially involves denying that the objects entailed by T_1 exist, then what kind of identity is established between the statement in T_1 and its correlate in T_2, if the terms used in X do not refer, while those in Y do?

C-rules establish an identity between predicates of one domain and predicates of another. Furthermore, the meaning of a concept is tied to its use. And since explaining the observational constitutes one function of the theoretical domain, then one of the two items correlated has a use which the other lacks. The function in question is that of explaining the empirical domain. One inference can be drawn from the theoretical level which cannot be drawn from the empirical, namely that the objects to which the empirical predicates pertain do not exist.

If this is the case, how can they both be said to have the same meaning? If they don't have the same meaning, they cannot be identical. If they are not identical, then *what* role does the C-rule play?

Problems over the interpretation of C-rules as identity statements stem from too simple a theory of meaning. While *starting* from the idea that the meaning of a word is its use, Sellars distinguishes five different uses. I do not intend to try, even briefly, to explicate Sellars' theory of meaning. A brief account could not possibly do it justice. It should suffice to know that one of the uses which Sellars distinguishes is:

the sense of 'expresses the concept . . . '. In this case we must say not
  (2) (The English adjective) 'round' means circular but
  (1) (The English adjective) 'round' means circularity or, as I shall put it,

(7) (The English adjective) 'round' expresses the concept circularity.
Notice that it would be incorrect to put this by saying that
(8) 'Round' names the concept Circularity for this is done by 'roundness'. ([44],
p. 111)

What the word does when it expresses the concept is to give its sense. This is not to say, however, that C-rules establish an identity of sense. For if this were claimed we would still run up against the problem of justifying the proposed identity. For what is the criterion by which we determine that the two phrases express the same concept?

C-rules, according to Sellars, express "more than a factual equivalence but less than an identity of sense". ([44], p. 125) They provide the means whereby we redefine observation terms in the language of the theory. But, again, this does not entail the elimination of the meaning of terms of the observation framework. For clearly, we then would view C-rules as actually implemented redefinitions. All Sellars seeks here is recognition of the role of C-rules, not their immediate success as tools for replacing the manifest image.

... for if correspondence rules cannot be regarded as implemented redefinitions, can they not be regarded as statements to the effect that certain redefinitions of observation terms would be in principle acceptable. ([44], p. 125)

The motivation for construing C-rules in this way should be clear. C-rules provide the basic mechanism whereby theories come to explain the perceptible objects common to the manifest image. They provide the basis for an explanation by putting us in a position to fully account for those objects by telling what they *really are* in theoretical terms.

6. EXPLANATION

A theory has two explanatory roles. First it explains phenomena referred to in another theory. Second, we use theories to describe and reason about the world. Let us recast this two-fold characterization in the following way: there are two types of explanation, internal and external.

The distinction between internal and external explanation is parasitic on Carnap's distinction between internal and external questions concerning frameworks.

If someone wishes to speak in his language about a new kind of entities, he has to introduce a system of new ways of speaking, subject to new rules; we shall call this procedure the construction of a linguistic *framework* for the new entities in question.

And now we must distinguish two kinds of questions of existence: first, questions of the existence of certain entities of the new kind *within the framework*, we call them *internal questions*; and second, questions concerning the existence or reality *of the system of entities as a whole*, called *external questions*. ([1], p. 206)

Sellars modifies Carnap's point that new entities require new rules. He distinguishes between (a) describing the behavior of the new entities within the framework, internal explanation, and (b) the *explanation* of the new entities. The latter involves defining the entities in terms of another framework, a theory, *T*.

Internal explanation occurs in the context of using a theory to reason about the world. It does not however produce *explanation*. Rather it shows the consequences of using a given theory. Given T_1, the behavior of some phenomena can be systematically described. This not only entails formulating generalizations with respect to its behavior, but also relating its behavior to the behavior of other phenomena. But this amounts to only a description, not an explanation. To have the so-and-so *explained* we must have recourse to another theory, T_2, which tells us what the so-and-so *is* by *redefining* it. That explanation requires redefining a concept in terms of a different framework is an obvious amplification of the general conflict Sellars sees existing between science and common sense. Science is to replace common sense by redefining its concepts. It does this in order to explain what common sense cannot. What we have in *SE*, therefore, is an extension of the discussion of Chapter I. To fully explicate this involves expanding the earlier formulation of *SE*. So let us replace the protothesis with the following.

> *SE Thesis*:   Explanation involves both the defining of terms in one framework, T_1, in the terms of another framework, T_2, using C-rules, and the denial of the reality of the objects denoted by the terms in T_1.
>
> (a) To explain x, which occurs in T_1, requires: both defining x in the vocabulary of T_2, and a rule to the effect that while whatever it is that is denoted by the *definition* of x in T_2 exists, the objects denoted by x, *as it occurs* in T_1, do not exist.
>
> (b) To define 'x' in T_2: use a substantive C-rule to correlate x with a statement in T_2, e.g., y, which has the same function in T_2 which x has in T_1.

SE incorporates an essentialist view of explanation. Explaining a phenomenon entails telling what it is. This contrasts with the major Positivist account Hempel's Deductive–Nomological Theory of Explanation, *DN*.

Hempel's Thesis: To explain an event described as *x*, show that *x* is deducible from a set of statements, at least one of which is a highly confirmed law with empirical content, the other true statements of antecedent conditions.

The *apparent* dissimilarities between the two views are glaring. While for Hempel explanation involves deducing a description (or law) from an empirically significant law, on Sellars' account it involves neither laws nor deduction. On the other hand, Sellars insists that explanation entails replacing descriptions in one theory by descriptions in another, while Hempel does not.

These surface differences, however, belie a more important set of similarities and differences. Both are realists. Both consider the role of theories crucially important in explicating that role. Hempel views theories as the key instrument in systematizing knowledge. ([16], pp. 179–182) Sellars, on the other hand, considers theories in their conceptual role of providing the framework for explanation, where that amounts to telling us what the item in question really is.

Are these more than merely verbal differences? For surely even Sellars' analysis of explanation entails systematization. There *is* more at issue than emerges on only a casual review of their explanatory theses. The major differences concern the place of induction in systematization.

Hempel keeps a special place within his theoretical structure for claims based on inductive reasoning. Not all knowledge forms a neat deductive system since some of our knowledge consists of statistical generalizations and probability statements. Hence we need to consider two different types of systematization, deductive and inductive. ([16], pp. 174–175)

Sellars, however, is a strict deductivist. He believes it is a mistake to give special status to a set of claims because of the manner in which they were obtained. Thus, for Sellars, a generalization is either a generalization or not. Calling it an *inductive* generalization confuses the reasoning surrounding the formulation of the claim and the content of the claim. ([40], pp. 270–271)

Despite the fact that the major differences between Hempel and Sellars concern induction, this issue lies hidden behind their apparent disagreements on the nature of explanation and the structure of a theory when, in fact, they do not differ. For, as it turns out, *SE* is not an explication of 'explanation', but rather a method for replacing theories in a general program which views the end of scientific inquiry as a final general theory.

DN, on the other hand, is compatible with what we earlier called internal explanation. Internal explanation concerns the means of reasoning about items within a theory using rules of inference. And since, as we have seen, Sellars analyzes laws as rules of inference, what Hempel calls explanation Sellars considers the basis for sound reasoning and at this point the differences are not crucial.

Hempel and Sellars also *apparently* disagree on the structure of a theory. Hempel *seems* committed to the levels view of theories which postulates a NL_O against which theories can be tested. Since Sellars argues that one is not committed to such an absolute level it would appear that Hempel has problems. But, Hempel does *not* insist on a NL_O. He weakens the characterization of the observation language to include whatever is antecedently clear in its meaning and accepted as playing an observational role by the general scientific community. But this is still unsatisfactory since, as I argue, Hempel's idea of antecedently clear presupposes the theoretical framework which it serves to help test.

Sellars objects to the Positivist NL_O for two reasons. First, he doesn't believe it exists. Secondly, the insistence on an NL_O coupled with the idea that theories are constructed to explain empirical generalizations results in the conclusion that theories are dispensable. Hempel has characterized this problem as the Theoretician's Dilemma. And, moreover, Hempel agrees that restricting science to only deductive systematization results in the dilemma. But Hempel argues that we must also include inductive systematization. Once we recognize this we see that theories are not dispensable.

Hempel, however, only succeeds in escaping the dilemma at the expense of placing inductive inferences in a special category. ([16], pp. 175, 222) And, as noted above, this confuses the method of arriving at a conclusion with the content of what is concluded. Moreover, and for our purposes, this constitutes the major point, by restricting inductive inferences in this way Hempel subverts the proper role of induction in science. By limiting discussion of induction to systematizing statements based on inductive reasoning, Hempel, and other Positivists, once again exhibit their concern with the product of science. But, change in science is our problem. Specifically, how do we rationally proceed from a theory we know to be false to one we assume to be true? We can only do so if we assume that inferring from what we know to what we do not know is rational.

7. ONTOLOGICAL PRELIMINARIES

We noted in passing that Sellars is a realist. *SE* lays bare this aspect of his theory of science. Not only are we to correlate x with y, but we must also rule that the entities x denotes do not exist. Sellars would have us 'rule from above'; that is, the more highly developed science would have the final word as to what exists and what does not exist. ([46], p. 353, and [45], pp. 32–37) *DN* entails no such ontological intervention, even though Hempel also happens to be a realist. Hempel agrees with Sellars that whatever science says is real we must assume to be. Moreover, like Sellars, he argues for the necessity of assuming a realist position if we are to make any progress in scientific inquiry. But, in general, *DN* leaves the problems of ontological commitment aside.

DN assumes that whatever else is the case, the common objects of observation, the facts to be explained, are real. It would be close to madness to deny the existence of the chair in front of you if the proper observational circumstances are operative (i.e., you are not observationally incapacitated, the lights bright enough to make observation possible, etc.). This general outlook can be weakened to the simple claim that, if anything, that which we see as objects must be real. And, moreover, we can adopt this weaker position without modifying the essential *DN* bias in favor of the observational context.

More important to *DN* than the commitment to the reality of the framework of common sense is its *lack* of importance in the logic of explanation. However the terms used in observation statements are interpreted (concerning the reality of their referents), the mode of explaining these statements remains the same. We can discuss the ontological status of the objects, facts, events, or phenomena in other contexts. In the same vein, the ontological status of the entities denoted by the theoretical terms of the theory has no essential bearing on the use of those terms in explanations.

On *SE*, by contrast, we are immediately forced into an ontological position and problems. If Sellars argues both (1) that there is no absolute F_o and (2) that the logic of explanation essentially involves the replacement of theories, he would appear to be involved in an infinite regress of replacing frameworks. This does not necessarily constitute an unacceptable result, especially if you do not view the universe as stable and unchanging. But that there is a determinate and knowable structure to the universe constitutes an essential Sellarsian presupposition. ([45], p. 21) Given this presupposition, the character of internal explanation becomes more important as science comes closer to completion. In the complete science we no longer need to inquire into the nature of the objects whose behavior we describe. The complete science accurately describes all aspects of the universe such that predictions are never falsified and it pictures the world as it really is.

It was noted above that the distinction between internal and external explanation is parasitic on the Carnap distinction between internal and external questions concerning the reality of entities and the nature of conceptual frameworks. Carnap makes a simple Kantian and pragmatic point. The ability to critically analyze the primitive concepts of a linguistic framework is limited when the analysis occurs within that same framework. A complete analysis requires another framework. For internal questions this could be a meta-language. When external questions are at issue we require a totally different framework. External questions in turn are decided on pragmatic grounds which are a function of the importance of the decision involved.

An essential aspect of *SE* is the determination of the ontological status of the entities described. Sellars thereby up-ends Carnap's distinction between internal and external questions by turning a condition of adequacy for answers to internal questions into an approach to solving the ontological problems characterizing external questions. For long-run purposes the ontological issues take precedence. An object is not explained until you know what it is. For example, we may describe a gas as a collection of molecules obeying certain laws, but molecules, not gases, are the real entities in the world. This is *not* a decisional matter at all, even though the choice of which theory to use in the end may be.

That the ontological status of the entities must be established in order to have a complete explanation may or may not be a worthwhile requirement, depending on what view we take of how science as an ongoing activity should be conducted. This is not to be confused with the issue of how to reconstruct a scientific theory to illustrate its logic. On one side, the irrelevance of the ontological status of the entities can be maintained. For the scientist, the important issue remains the conducting of experiments and the postulation and testing of hypotheses which at some later point can be systematically collated to form a theory. The theory should explain the phenomena observed in the world, no matter what they really *are*. But, if that latter point continues to be bothersome, then after we construct the theory we can discuss its ontology.

As an example of the alternative position, for the methodological importance of determining the ontological status of the entities of a theory, Paul

Feyerabend's views stand out. Unless the scientist argues that the entities postulated by his theory are real, Feyerabend contends, the possibility of working out the theory in its best form diminishes. ([8]) Insisting on the reality of those entities compels the scientist to struggle to save a promising theory in the face of what would normally, in *DN*, for example, constitute disconfirming evidence.

Feyerabend's pragmatic point of view concerning the manner in which satisfactory and useful theories can be produced in science is again a variant on the theme of collapsing the distinction between ontology and epistemology. The Positivist position attempts to retain the distinction. There it turns into the logical distinction between the logic of explanation and the content of a theory. Sellars uses neither the pragmatic nor the Positivist point of view in this case. Instead he argues for a position which, like Pierce, assumes a linear progression in science culminating in one theory. Against Feyerabend, Sellars contends that you can't be ontologically both a realist and a pragmatist. If we are ever to construct theories which are true, realism is necessary, not merely useful. ([46], p. 353)

If (a) the universe is only fully explained when we know of what it is composed, and (b) if knowledge presupposes a conceptual scheme which determines (i) the form of knowledge, (ii) the content of knowledge via the categories of the scheme, and (iii) the criterion of truth as semantically assertible within a theory, then (c) knowledge of the ontological status of an entity requires the use of a conceptual scheme. But to know fully entails knowing the ontological status of the entities, which in turn requires another framework. If a conceptual scheme becomes the final arbiter of reality, and if knowing how to use the conceptual scheme constitutes the essential dimension of knowledge, we eliminate the chance to determine both the truth of any given scheme and the reality of the objects it postulates.

8. EXPLANATION AND EXISTENCE

A theory is an explanatory framework, $T = F_e$. The legitimacy of this equivcation between 'theory' and 'explanatory framework' presupposes Sellars' complex description of a theory as a formal deductive system embedded in a meta-language of rules concerning the use of the formal system. *A framework is a context in which actions of specified types can be accomplished.* In a theoretical framework, F_t, statements involving theoretical concepts are formulated and used. In an observation framework, F_o, observation claims occur. Obviously this characterization of a framework also entails that the conceptual scheme available to agents who use it to organize data constitutes one of the defining characteristics of any given framework.

Another defining characteristic of a framework is the set of rules which determine how claims formulated within the framework are to be used. Claims formulated within F_t are used in explanations. But one cannot offer *explanations* outside the context of a *theory*. In those cases, i.e., from a Positivist point of view, where a formal deductive system exhausts the exhausts the analysis of a theory, there exists a wide difference between T and F_e. But the distinction collapses in cases such as Sellars', where the analysis of a theory results not only in a complex of interrelations between an object language and a meta-language, but where a specific set of rules in the meta-language concerns the role of a theory: explanation.

If no absolute F_o exists, then any F in which empirical generalizations and observation reports can be formulated can qualify as an F_o. This is especially true once we weaken NL_o to accord with Hempel's criterion of accepting whatever the community at the time accepts as an observation term. We are then left with merely an observation language, L_o. Particularly relevant to our discussion here is Hempel's inclusion in L_o of terminological fallout from old and sometimes even abandoned theories. It is then possible for some theory, T_1 to function as a part of an F_o. Those terms form part of the language, L_o, used in F_o to make observations. In such cases we classify the predicates describing entities as empirical.

Assuming for the moment a distinction between explanation and derivation, we can discuss the two stage process of explanation proper. First, using C-rules we correlate one theory, T_1, with another, T_2. This means the objects of T_1 can now be explained by T_2 where $T_2 = F_e^2$. Since it is not deductive we distinguish this process from the derivation of statements within a theory, an aspect of internal explanation.

Once we correlate T_2 with T_1 the explanation of the 'object' or 'event' in question in T_1 can proceed. The correlation which the C-rule establishes provides the grounds for replacing the empirical account of T_1 by the theoretical of T_2. The event is then properly located in the theoretical domain and the explanation can continue to the second stage: the elimination of the empirical description in T_1 in favor of the theoretical in T_2.

Several problems arise when we put together the account sketched above and the view "that in principle it is the framework of theory *rather than* the observation framework which is real". ([46], p. 359) For example, it is indeed strange to say both that only the framework of theory, F_t, is real and that F_t explains F_o. The oddness rests on two points. First, we have it that a C-rule establishes the following *as an explanation*: x, in the theoretical domain, is correlated with y, in the observation domain, and the explanation of y is that it doesn't exist, x does. Second, while it may be reasonable to say the objects of F_e^1 as opposed to F_e^2 are real, the claim that the *framework* is real doesn't make sense. Perhaps it could be explained, but it would lead us far afield into metaphysical questions concerning the nature of language. Therefore let us assume that Sellars here means the objects denoted by the language of a given explanatory framework, F_e, and *not* the framework itself when he speaks of the reality of things.

Now, in what sense does the claim that something doesn't exist constitute an explanation?

According to the view I am proposing, correspondence rules would appear in the material mode as statements to the effect that the objects of the observational framework *do not really exist – there really are no such things*. They envisage the *abandonment* of a sense and its denotation. ([44], p. 126)

It is quite unintuitive to say that a C-rule correlates objects in different domains, and that one of these domains is empty. How can the theoretical account replace a description which has no referent, and then itself refer?

One further problem should be mentioned at this point. The mere matching of observational items with their theoretical counterparts *by itself* constitutes no explanation. That remains merely the replacing of one set of descriptions by another where, Sellars claims, the former has neither sense nor denotation. The adequacy of the description must rest on something other than its theoreticalness. There must be something more to explanation for Sellars than the mere replacing of frameworks.

9. EXPLANATION AND TWO SENSES OF 'ABOUT'

We noted earlier that theories are explanatory frameworks. This is to say they are *about* things. There are, however, two senses in which theories are about things. First, a theory, T, describes the objects it mentions and offers rules for reasoning about the behavior of those objects. Second, T is 'about' something in the sense of explaining that thing. The former amounts to internal explanation, the latter external explanation. The distinction at work here can be elaborated by considering the difference between a theoretical framework, F_t, and the framework of common sense, F_{cs}, in terms of their subject matter.

In the context of arguing (against Feyerabend) for a distinction between

F_t and F_{cs}, Sellars observes that claiming common sense objects constitute the subject-matter of F_{cs} trivializes the issue. ([46], pp. 338–339) In one sense every T has a similar domain, i.e., its equivalent of common sense objects. Every T has an *internal* subject matter; the cash value of the first of the two senses of 'about' noted above. In the case of the theory of gases the internal subject matter is molecules and the behavior of molecules. But, a theory also has an *external* subject matter, that which it explains, e.g., gases. This differentiates T from F_{cs}. Theories are related to their external subject matter by C-rules. The difference then between some F_t and F_{cs} is that "the conceptual framework of common sense has no *external* subject matter and is not, therefore, in the relevant sense a theory *of* anything". ([46], p. 339)

Even though theories *use* concepts according to rules, the role these concepts play within T is not explanatory with regard to the objects named. Rather, the functioning of the concept illustrates the behavior of the objects so far as that behavior relates to other significant aspects of T. Thus F_{cs} provides the opportunity for *describing* the world but not explaining it because the descriptions of the world in F_{cs} do not replace descriptions phrased in another F. To explain an object requires, in Sellars' sense, the introduction of new entities and rules, which, in effect, constitutes the creation of a new F_e with wider scope than the old one.

So far then, on SE, theories are frameworks which have the characteristic of being explanatory because they have an external subject matter which they are about. However, it is not obvious that being about something explains that thing. For example, thoughts are, whatever else, sometimes about things, and moreover, thoughts and things belong to different frameworks. But if one has a thought of an object the object is not somehow explained. Clearly this cannot be all Sellars has in mind and in responding he claims the logic of labeling a series of sentences a 'theory' involves the associated concept of explanation. The two notions are in the same 'logical neighborhood'.

This may very well be true, but the use of Sellarese leads to many problems in understanding the content of such remarks. As a first approximation, we can say that by calling a set of sentences a theory, T_1, we mean T_1 explains some other set of sentences, T_2. This is accomplished *in part* by devising a set of rules (C-rules) for replacing a sentence in T_1 by one in T_2. The internal subject matter of T_1 constitutes the external subject matter of T_2. In the case of microtheories, we correlate descriptions of objects of diffferent types, collections of microentities with macroentities. But a correlation is not a deduction, hence on DN not an explanation.

10. EXPLANATION VERSUS DERIVATION

Sellars distinguishes explanation from derivation because confusing the two results in the rejection of theories as useful in science. He credits the position he opposes with postulating three different levels of development. (While *DN* lends itself to being described in this way it does not follow that proponents of *DN* fail to distinguish these concepts. See for example, Hempel's distinction between explanatory relevance and testability in [16], pp. 47–49. Mere derivability while a necessary condition is not a sufficient condition for an explanation.) According to *DN* the bottommost level is absolute in some sense, and, in Sellarese, constitutes the level of Explained Nonexplainers. ([44], p. 120) These are observation statements. They are explained by the statements of the second level, inductively based empirical laws, the Explained Explainers. These in turn are explained by the top level of unexplained Explainers, or theoretical principles. Describing the motivation for this type of analysis, Sellars (using 'we' as the common sense we) notes that,

> The more important source of the plausibility of the *levels* picture is the fact that we not only explain *singular matters of empirical fact* in terms of *empirical generalizations*; we also, or so it seems, explain these generalizations themselves by means of *theories*. [44], p. 120)

He observes that the initial plausibility of this account can be misleading if either of two things are the case. First, if "it finds too simple . . . a connection between explaining an explanadum and finding a defensible general proposition under which it can either be subsumed or from which it can be derived with or without the use of correspondence rules". ([44], p. 120) This is a caution against collapsing the distinction between explanation and derivation. To put Sellars' cards on the table, recognize from the beginning that an adequate explanation of an event must provide not only reasons, in the form of laws which describe the behavior of the object, but it also must account for why the event in question would *not always* obey an inductively based generalization.

According to *SE*, collapsing the distinction between derivation and explanation precludes the possibility of *explaining* both (a) the derivability of observation statements from generalizations and (b) the *failure* of a prediction arrived at by the derivation of an observation claim from a generalization. ([44], p. 121) That is, we can only account for the failure of the prediction if we can give an account of the objects mentioned in the observation statement such that it explains why they behave the way they do, i.e., why they sometimes obey the generalization and why they sometimes don't. This requires (a) a description of the object in terms of a different T, i.e., T_2 (given that we have exhausted the definitional apparatus of the theory in use, T_1) *via* C-rules and (b) rules of inference guiding the role of the defining concepts in T_2 so that it can be used to illustrate the structure of the world (in its own terms, of course).

The issue turns on Sellars' analysis of empirical generalizations. While on *DN* no completely adequate analysis of a law yet exists, there is agreement that empirical laws, laws containing empirical predicates, must be essentially general and true. According to Sellars, the empirical generalizations *DN* uses in derivations and explanations of observation statements cannot be lawlike, and, hence, cannot function as the nomological elements in an explanans. There are two reasons for this claim. First, empirical generalizations are inductively based and, according to Sellars, are all to be properly characterized probabilistically (universal empirical generalizations are limiting cases) since the evidence for them provides only good reasons for using them and does not conclusively determine their truth.[2] Second, Sellars treats laws as rules of inference. Generalizations based on induction *presuppose* a framework of such rules. On this point Sellars differs radically from Hempel. Hempel views systematization of inductively based claims as part of the scientific enterprise. Sellars claims science presupposes an inductive or ampliative framework.

Thus, according to Sellars, empirical generalizations cannot function as explained explainers if this entails that observation statements be derivable from them. They are based on induction and as such are unjustifiable as *universal* claims. An 'explainer' must be universal. Sellarsian rules of inference, on the other hand, function in a manner analogous to 'ought' statements. They tell one what the world ought look like. Empirical generalizations are formulated in the context of these rules and hence cannot have the same explanatory role. That is, laws of nature justify using empirical generalizations because they are rules for using them. Generalizations, as probability claims, cannot do the same thing.

The proper role for empirical generalizations lies, as we saw in Chapter III, in accepting theories. Briefly, if T_1 explains the observation statements by more fully describing the objects (by relating their behavior to other newly postulated entities) and by permitting the same basic observations to still be significantly made, then it should indirectly explain the generalizations based on the observation claims. Because the objects are of a certain kind which behave in a manner described by T_1, we can explain the probability of a given generalization. The higher the probability of the generalizations the better the theory.

There is also a second point to be guarded against in considering the levels theory of explanations.

... it is supposed that whereas in the observation framework inductive generalizations serve as principles of explanation for particular matters of fact, microtheoretical principles and principles of explanation *not* (*directly*) *for particular matter of fact in the observation framework, but for the inductive generalizations in this framework* (*the explaining being equated with deriving the latter from the former*) *which in their turn serve as principles of explanation for particular matters of fact.* ([44], p. 120)

Theories do not first explain generalizations which in turn explain particular matters of fact. Theories directly explain particular matters of fact. Forgetting this point can result in a crucial mistake,

for to conceive of the *explananda* of theories as, simply *empirical laws* and to *equate* theoretical explanation with the derivation of empirical laws from theoretical postulates by means of logic, mathematics, and correspondence rules is to sever the vital tie between theoretical principles and particular matters of fact in the framework of observations. Indeed, the idea that the aim of theories is to explain *not* particular matters of fact *but rather* inductive generalizations is nothing more nor less than the idea that theories are in principle dispensable. ([44], pp. 120–121)

There are two points worth considering here: (i) the issue of explaining laws, and (ii) the problem of understanding why if theories explain inductive generalizations this leads to dispensing with theories.

(i) In the first case Sellars believes that if deriving a law from a theory counts as explaining the law, it is impossible to explain particular matters of fact. This holds both for when a law is either the logical product of derivations from the postulates of T or an inductively based empirical generalization. Thus, if we derive the law from the postulates of T, where T meets Nagel's characterization, then the problem arises of insuring the relevance of any such theoretical formulation to some observation statement. This problem emerges in either of two cases. (a) if the postulates of T are derived from an uninterpreted T_c then the relation between a schema, i.e., a postulate, and some semantically meaningful statement is not obvious, and a claim that the derivation of that statement from the schema constitutes an explanation is subject to considerable doubt. (b) With an interpreted calculus, where the theoretical terms are only partially defined (which is the best DN can do if they are not to be eliminated completely from the theory; see discussion of Theoretician's Dilemma below), then the sense in which a statement using *partially* defined notions can *explain* another statement remains extremely dubious where the criterion of an adequate explanation requires complete answers.

Where the law is the product of inductive reasoning over particular observations, there are three objections. (a) Induction presupposes a framework of rules which govern the formation of generalizations on the basis of individual observations. (b) Inductive generalizations cannot fully account for the behavior of the objects mentioned in the observation statements. The products of induction, they must be properly viewed as probability claims about the degree of support the evidence upon which they are based provides for a more general claim. Thus, at best, they can only provide a probable explanation for the observation claim which, as we saw earlier comes to no explanation at all. (c) Inductive generalizations play a role in justifying the acceptance of lawlike statements and are not to be confused with those statements. Lawlike statements tell us what the world ought to look like. Probability statements tell us that there are good reasons for accepting the claim that the world ought to look that way.

(ii) In order to see why Sellars claims that adhering to the levels picture of theories commits us to the position that theories are in principle dispensable, let us turn to Hempel's discussion of the same issue in 'The Theoretician's Dilemma'.

11. THE THEORETICIAN'S DILEMMA AND THE LEVELS THEORY OF THEORIES

The dilemma can be stated briefly: (a) since the purpose of a scientific theory is to establish connections among observables, and (b) since any theory which accomplishes this can be replaced by a law which directly connects its observational antecedents and consequents, then (c) if a theory accomplishes its purpose, it can be eliminated (in favor of the law phrased in terms of observables), and if it doesn't, then it is obviously unsatisfactory and can be eliminated. Thus, no matter which way we turn, theories are eliminable.

In his own account Hempel supports Sellars' critical description of the levels view of theories.

... it will be helpful to refer to the familiar rough distinction between two levels of scientific systematization: the level of *empirical generalization*, and the level of *theory formation*. The early stages in the development of a scientific discipline usually belong to the former level, which is characterized by the search for laws (of universal or statistical form) which establish connections among the directly observable aspects of the matter under study. The more advanced stages belong to the second level, where research is aimed at comprehensive laws, in terms of hypothetical entities, which will account for the uniformities established on the first level. ([16], p. 178)

While Hempel has only noted two levels here, the third level is implicit in his account since generalizations are formed on the basis of observation reports and are concerned to establish regularities. He also distinguishes two classes of vocabulary involved in formulating statements on these levels, observation terms and theoretical terms, with the latter functioning only in statements at the most 'advanced' stage.

Second, in characterizing theoretical terms, Hempel also makes one of the two moves against which we were warned by Sellars: "they function, in a manner soon to be examined more closely, *in scientific theories intended to explain empirical generalizations*". ([16], p. 179, italics mine) Hempel describes the purpose of a theory in terms of explaining generalizations, rather than matters of fact, the very point Sellars emphasizes should *not* be made. Hempel, however, does modify this characterization in his solution to the Theoretician's Dilemma.[3]

Finally, earlier we called the postulation of levels an aspect of the Positivist 'mythology'. That it does not describe science as it actually proceeds constitutes the grounds for this charge. In reply it can be argued that it does not pretend to be concerned with science as it historically develops. It concerns the rational reconstruction of science. Putting aside questions of what a rational reconstruction comes to, it is still hard to explain the sort of thing Hempel says, for instance, in the long section quoted above. There he talks about the *development* of a scientific discipline, and offers what appears to be a description of theory construction *as it actually occurs*; hence, the talk about the 'early stages', 'development', and 'more advanced stages'. Thus, it seems appropriate to level a criticism against Hempel here which also applies to Kuhn. ([23]) It is not clear whether Hempel offers a description of science as it actually develops, or a framework in which to discuss theory construction, as it is often not clear which of the two Kuhn offers. A charitable and not unreasonable account of Kuhn's work suggests be uses the history of science to develop a theory about both (a) the relationship between theory and the world, and (b) the nature of scientific change which serves as a model for reconsidering the hitherto unquestioned supremacy of the role of evidence in positivistic theories of science.

Where Kuhn at least falls back on the history of science to support some of his claims, Hempel does not, perhaps because he does not intend his account to be historically correct. I do not intend to argue the falsity of Hempel's account of the development of theories. If pressed, Hempel would probably claim that what he offers as a description constitutes only a very schematic version of the history of science, in much the same way as Sellars' account in 'Philosophy and the Scientific Image of Man' remains only a schema for discussing the development of science.

It is also worthwhile remembering that for Sellars the systematization of knowledge is essentially deductively organized once obtained, regardless of the type of knowledge claim. Induction, on the other hand, plays a crucial role in both theory building and theory acceptance. It also plays a crucial role in the justification of the use of a deductively systematized body of information as well as in its development. For Sellars the very possibility of theory construction presupposes induction. Theory construction proceeds in an ampliative framework. That we may systematize whatever knowledge we may have inductively arrived at is not a point of contention. But Sellars does challenge the view that once systematized the product should continue to be considered inductive knowledge. The probability statements which are the product of induction can be used only in the context of a system of practical reason. And Sellars' system of practical reason is deductive. Probability statements, once systematized, become part of the object-language of a theory. The rules of the meta-language contain the system of practical reasoning. We can therefore reason deductively about probability claims.

At this point it may be possible to shed some light on the process-product analysis of science. As mentioned earlier, Sellars attempts to develop an account of science that will handle not only the product of scientific inquiry, i.e., theories, but also one which can explain the development of science. When Hempel correctly notes that not all knowledge occurs in the form of universal claims, i.e., some is probabilistic, Sellars does not deny it. Instead, he denies the subsequent inference that because some statements have less than universal scope deductive reasoning does not govern their use.

If science uses only one type of systematization, then isn't Sellars also faced with the Theoretician's Dilemma? Hempel escapes only by including inductive systematization in his analysis of science. The answer here again is 'no' because Sellars liberalizes earlier verificationist principles by introducing some members of the theoretical vocabulary of a theory, T_{vt}, as primitives. Members of the observational vocabulary of a theory, T_{vo}, derive their meaning from the role of the theoretical terms and *not* vice versa. But, just as Hempel has to justify using theoretical terms as defined, Sellars also needs to justify his use of these terms as primitives. Sellars' justification involves two different points.

The first point involves several different claims. (a) Science tells us what exists, in the same neo-Carnapian way that the existents of a linguistic framework are determined by the rules of that framework. (b) The following corollary to Sellars' use of C-rules: the statement in F_t must be capable of eventually taking on the observation role that the explained statement in F_o currently has. Thus, if T_k completely replaces F_o, then the entities it postulates must be capable of being observed.

Secondly, there is the heuristic role of a model in the construction of alternative and new theories. If we first construct a model and argue for the entities of a new theory by analogy with the parts of the model which correspond to those entities, we can introduce those theoretical notions as primitives, and yet not uninterpreted. They will be, in Hempel's phrase, antecedently clear.

12. SELLARSIAN SYSTEMATIZATION

Sellars' counter to Hempel then is to recast the notion of systematizing knowledge. Hempel, a firm adherent to the levels view, extricates himself from the Theoretician's Dilemma by assuming that science involves both inductive and deductive systematization. Hence, Sellars may still be correct if he can cogently deny the justifiability of inductive explanation and prediction. To do so, however, we require more than simply Sellars' proposal of a different way to view the role of inductive generalizations. For, as he himself notes, the levels view and its corollary that we explain matters of fact by appeal to empirical generalizations and these generalizations by appeal to theories have a certain plausibility.

The 'something more' we require is a reason to reject the use of inductive generalizations as explanatory and predictive tools. This reason can be located in Sellars' explication of theoretical explanation. In his attack on the levels theory of explanation he takes a misleading first step. He appears to be concerned with the levels theory only to the degree that it involves an absolute L_o. The real issue, however, centers on the role of inductively based empirical laws, the explained explainers.

Generalizations do not explain particular matters of fact, theories do. In fact, theories,

explain empirical laws by explaining why observable things obey, to the extent that they do, these empirical laws; that is, they explain why individual objects of various kinds and in various circumstances in the observation framework behave in those ways in which it has been inductively established that they do behave. Roughly, it is because a gas is — in some sense of 'is' — a cloud of molecules which are behaving in certain theoretically defined ways, that it obeys the *empirical* Boyle-Charles law. ([44], p. 121, italics mine.)

Now if we explain laws by describing both the entities they concern and their behavior, then Sellars seems justified in his next move:

Furthermore, theories not only explain why observable things obey certain laws, they also explain why in certain respects their behaviour obeys no inductively confirmable generalization in the observation framework. ([44], p. 121)

Thus, by explaining why objects behave the way they do we can also explain why a given generalization has a certain probability.

Explanations are found in the establishing of C-rule relations between some T and the F_O in which the phenomena to be explained are described. Given T, the reason is forthcoming why the objects of F_O are *observed* to behave in some random fashion, i.e., in a manner described in observation terms only by a probability statement. The objects in question behave sometimes in this fashion and sometimes fail to behave that way because they are really configurations of ys which are objects whose behavior is described in F_O observation terms. This often involves elaborating T so as to accomodate distinct pecularities.

For example: given two observationally identical specimens of gold, if when placed in *aqua regia* they are observed to dissolve at different rates, we first explain this by modifying the theory.

The microtheory of chemical reactions current at that time might admit of a simple modification to the effect that there are two structures of microentities each of which 'corresponds' to gold as an observational construct, but such that pure samples of one dissolve, under given conditions of pressure, temperature, etc., at a different rate from samples of the other. ([44], p. 121)

By modifying the theory in this fashion, by postulating two entities rather than one, we explain the observational anomaly. The mark of success for the explanation is its coherence with the theory's general account of chemical reaction, i.e. no ad hocing permitted.

Thus, microtheories not only explain why observational constructs obey inductive generalizations, they explain what, as far as the observational framework is concerned, is a random component in their behaviour, and, in the last analysis it is by doing the latter that microtheories establish their character as indispensible elements of scientific explanation. ([44], p. 122)

The nature of explanation, as Sellars sees it, not only provides a means of introducing modifications in the theory doing the explaining, but also accounts for the sense in which theories can criticize and justify changes within the observational framework. Thus,

> microtheories explain why inductive generalizations pertaining to a given domain and any refinement of them within the conceptual framework of the observation language are at best approximations to the truth. ([44], p. 123)

The sense in which theories can explain how inductive generalizations are approximations to the truth requires more than just showing why objects behave in ways that do not lend themselves to accurate description by universal statements formulated in the observational framework. This requires an understanding of the manner in which a theory can criticize these generalizations.

> Where microexplanation is called for, correct macroexplanation will turn out (to eyes sharpened by theoretical considerations) to be in terms of 'statistical' rather than strictly universal generalizations. But this is only the beginning of the story, for the distinctive feature of those domains where microexplanation is appropriate is that in an important sense such regularities as are available are not statistical *laws*, because they are unstable, and this instability is explained by the microtheory. ([44], p. 122)

Imagine a domain of idealized inductive generalizations about observables where the idealization amounts to eliminating errors of measurement or experiment. This permits attention to be focused on the logical-mathematical structure of the idealized statements. An explanation of these statements requires a statistical category in the microframework, a category in which we can coherently postulate the instability mirrored by the statistical statements.

Thus, only with a theory which accounts for the nature of the statistical elements can statistical statements be revised and criticized. As Sellars points out, this is a logical point. The availability of a theoretical account which 'explains' their nature, and, hence, acts as a justification for their manipulation, is a prerequisite for the systematization of statistical generalizations.

The following constitutes another way of putting this same point. Before the observational framework can yield any explanation, a theory must be operative which provides the logical basis for operating on statements formulated in that framework.

13. EXPLANATION AND EXISTENCE: THE ROLE OF C-RULES

We can now partially answer an earlier problem concerning the reality of F_t

and F_o. Explanation involves establishing what the objects in F_o *are* by using C-rules to correlate statements in F_o and F_t. The problem arises when we also claim that (a) the objects of F_t are the real ones, and (b) that the C-rule permits the abandoning of the sense and denotation of the names used to mention the objects in F_o. What sense are we to make of C-rules and the relation they establish when the names occurring in one of the two frameworks fail to denote?

For Sellars C-rules are definitional in character. We redefine the right-hand side in terms of the left-hand side, thereby redefining T_{vo} in terms of T_{vt}.

The force of the 'redefinition' must be such as to demand not only that the observation-sign design correlated with a given theoretical expression be syntactically interchangeable with the latter, *but that the latter be given the perceptual or observational role of the former so that the two expressions become synonymous by mutual readjustment*. ([44], pp. 125–126)

The 'force' Sellars refers to stems from the syntactical dimension of this move. This reduces the consequence of reading C-rules this way to merely acknowledging the in-principle acceptability of redefining T_{vo} without requiring further that T_{vt} immediately take over the perceptual role.

The merits of this approach are two. First, it provides a legitimate base for claiming that in time (once the theory has become sufficiently entrenched or accepted, or recognized as true, or however we may come to describe the phenomenon of using and not questioning the veracity of a theory) the theory *may* replace the observational framework, however remote the present possibility.[4]

Second, it provides a way out of the paradox which arises from describing objects in both theoretical and observational terms. How many objects are there really out there? Are there both chairs and electrons, or only chairs, or only electrons? By claiming C-rules establish a syntactical identity and recognizing that this succeeds in "expressing more than a factual equivalence but less than an identity of sense", ([44], p. 126) we explain away the paradox by showing that there is really one entity with two names. Relating the names through C-rules, we "explain how theoretical complexes can be unobservable, yet 'really' identical with observable things". ([44], p. 126)

The crux of Sellars' argument for using C-rules as the explanatory device *par excellance* rests on an ontological point. Sellars denies Hempel's view of the aim of science as systematization. Rather, the aim of science is to tell us what there is. Explaining the behavior of entities without explaining what those entities are accomplishes only half the job. If science is to tell us what there is, then if we develop T_2 and it offers a more comprehensive account for all the phenomena accounted for by T_1 by redefining T_1 concepts in T_2 terms, then not only should we accept T_2, but in so doing we are committed to the ontology of T_2, having originally been committed to the ontology of T_1.

Hempel also acknowledges this point. In fact he holds a position far closer to *SE* than either he or Sellars may realize. Consider the following observation:

If the sentences of a partially interpreted theory T' are granted the status of significant statements, they can be said to be either true or false. And then the question, ..., as to the factual reference of theoretical terms, can be dealt with in a quite straightforward manner: To assert that the terms of a given theory have factual reference, that the entities they purport to refer to actually exist, is tantamount to asserting that what the theory tells us is true; and this in turn is tantamount to asserting the theory. ([16], p. 219)

Accepting the theory yields a commitment to the entities of the theory. The commitment to the entities of the theory means theoretical statements are meaningful. But since it is one thing to accept a theory and another to show it is true, we must test the theory. Hempel's testing procedure closely resembles Sellars'. We cash out theoretical claims in terms of predictions formulated in some basic observational vocabulary (V_B for Hempel).

Thus understood, the existence of hypothetical entities with specified characteristics and interrelations, as assumed by a given theory, can be *examined inductively* in the same sense in which the truth of the theory itself can be examined, namely by empirical tests of its V_B-consequences. ([16], p. 220, italics mine.)

Assume the entities exist; perform experiments designed to test those assumptions, some of which may take place in F_o, i.e., the observation framework. Since everything described in F_o can be translated into F_t this poses no problem. But how do we inductively examine the truth of T? Do we tally the evidence from experiments on the one hand and see if they give more or less support to T? Can we form inductive generalizations on the basis of observation concepts, and see if postulating the entities mentioned by T explain those generalizations as Sellars claims? There seems to be nothing barring this possibility.

Even though Hempel's account here seems comparable with Sellars', their rules for accepting those generalizations differ. Hempel follows Carnap by considering such factors as the total evidence requirement and the use of epistemic utilities. Sellars uses the straight rule which he confines to the object language level of T and makes the acceptance of a generalization based on the use of the straight rule a function of the scheme of practical reasoning found in T's meta-level. Those problems have already been discussed.

The difficulty with Hempel's view lies in his reliance on a type of theory in which the theoretical terms are partially defined by means of an antecedently understood set of terms which are members of L_O.

Generally speaking, we might qualify a theoretical expression as intelligible or significant if it has been adequately explained in terms which we consider as antecedently understood. ([16], p. 218)

How are we to read 'antecedently understood'? If Hempel only requires that the set of terms used defining theoretical terms have been used before and that its use be understood, then this is insufficient. Consider his discussion of understanding theoretical concepts:

In defense of partial interpretation, on the other hand, it can be said that to understand an expression is to know how to use it, and in a formal reconstruction the 'how to' is expressed by means of rules. Partial interpretation as we have construed it provides such rules. ([16], p. 218)

Who or what provides the rules for the 'how to' which would account for the sense in which the antecedently understood terms of L_O are understood? And most importantly, even with some account of those rules forthcoming, what assures us that the terms of L_O are appropriate to the use for which they are put? What is the criterion for choosing the members of L_O? If we return to Hempel's analysis of an interpretive system we note that L_O operates as a base for an interpretive system only insofar as its members occur essentially in a set of sentences J. Briefly then, it would appear as if the terms chosen for membership in L_O are not only antecedently clear but their relevance for the role of providing an interpretation is *already* determined. But what would be the criterion for determining their relevance? Certainly not that they are merely observation concepts. For not all observation concepts are relevant to every theory. For example, color concepts are irrelevant to the testing of the theory of gases.

There are two possible alternatives. First, L_O contains every antecedently understood term available. But this is unreasonable since it would render meaningless the following condition: that J "contains every element of V_t and V_B essentially, i.e. is not logically equivalent to some set of sentences in which some term of V_t or V_B does not occur at all". ([15], p. 208) (Where $L_O = T_{vo} = V_B$ and $T_{vt} = V_T$) Effectively this means that if V_B

contains every meaningful term available, then no set of sentences logically equivalent to J can be formulated since every other sentence using a non-theoretical term must use a member of V_B.

The second alternative requires both that the members of L_O be antecedently understood, and that the members of T_{vt} be antecedently understood as well. With their use already determined at their introduction in T then they can determine the relevant set of terms necessary for L_O. But if so, the members of L_O will not be necessary for the purpose of interpreting the terms in T_{vt} since they will already be understood. They can, however, be used to help test T in accordance with procedures which require a type of observation using those concepts.

However, we are not yet out of the woods. For it is one thing to suggest that the members of T_{vt} be antecedently understood before admitted, and quite another to explain how they could be.

For Hempel the *interpreted* postulates incorporate the fundamental assumptions of a theory. What counts as an interpretation of the postulates lies at the source of the worry on how to introduce theoretical terms into a theory. Contrary to the road Hempel has taken, Sellars claim that,

The fundamental assumptions of a theory are usually developed not by constructing uninterpreted calculi which might correlate in the desired manner with observational discourse, but rather by attempting to find a *model*, i.e. to describe a domain of familiar objects behaving in familiar ways such that we can see how the phenomena to be explained would arise if they consisted of this sort of thing. ([41], p. 182)

The development of a theory does not proceed by first conducting haphazard experiments and then some monumental collation of facts and hypotheses into a deductive system. Rather, given the problem of explaining the behavior of some unknown phenomenon, the scientist looks to an area of relative similarity. In this area the objects are already specified and their activity explained. The similarity between the objects of this domain, F_t^1, and the one under examination, F_t^2, constitute the basis for developing a theory to explain the phenomena in question. We do not use the model, which consists of objects analogous to the objects of F_t^2 and the description of their behavior without qualification.

The essential thing about a model is that it is accompanied so to speak, by a commentary which *qualifies* or *limits* – but not precisely, nor in all respects – the analogy between the familiar object and the entities which are being introduced by the theory. ([41], p. 182)

The commentary makes explicit the 'as if' relationship between F_t^2 and the model. The objects in F_t^2 behave *as if* they were objects of the model, but there are differences which the commentary delineates. The descriptions of the behavior of restricted and modified objects of the model from the basis for formulating postulates of the new theory.

It is the descriptions of the fundamental ways in which the objects in the model domain, thus qualified, behave, which, transferred to the theoretical entities, correspond to the postulates of the logistical picture of theory construction. ([41], p. 182)

The criterion for determining which theoretical concepts are antecedently clear, and subsequently which ones are to be taken as defined, remains a different problem. Sellars does not address himself to this question, but the following can be extrapolated from his views. Those terms which retain their importance in earlier theories and which function in the more important laws of the theory seem to have *a priori* reason for being held antecedently clear, e.g., 'mass'. This position rests on Sellars' desire to trace the development of the key concepts once we have a finished science.

The view that old established terms have an edge on antecedent clarity constitutes a key element in Sellars' theory about the history of science. The concept of a linear progression in the development of the finished product holds the central position in this program. However, the assumption of a linear progression in the development of science remains dubious if not historically falsifiable. But given that assumption we can also assume that Sellars believes those concepts which science is supposed to refine are the very ones which paradoxically are taken as clear to begin with. The development and refinement of those concepts, therefore, consists in working out their logic.

The important point of this discussion is to show how using a model and a commentary does generally explain how theoretical concepts can be shown to be antecedently clear. We can also note the similarity between a model and its commentary and Sellars' description of a theory as having an object-level and a meta-level. Just as a model is useless without a commentary to explain its relation to other things, so too a deductive system is useless without the rules which tell people what they can do with the system.

It is apparent that Sellars provides an analysis of what Hempel means by 'antecedently clear'. The use of models is open to Hempel in just the way that it is open to Sellars. On the strength of such a device the relevance of L_o to the theory is easily marked out. Once we specify the meaning of the theoretical terms using the model, the relevant terms of the basic vocabulary can be picked to fill out the definition of those terms and test the theory.

If this analysis is correct it would appear that there are few, if any, major difficulties in principle between Sellars and Hempel on the issue of introducing theoretical terms. There are still difficulties on the issue of inductive systematization; and, of course, this is the crucial point.

Each of the attempts to eliminate the dilemma Hempel has considered he rejects, either because it avoids the issue by a sleight of hand, or because it lands us back in the dilemma, or because it fails to account for inductive systematization. And in his concluding comments Hempel feels that the dilemma only really arises because of the original attempt to limit scientific inquiry to deductive systematization. Hence, the real stumbling block is incorporating inductively based conclusions into a scientific theory.

But now if we turn our attention to the problem of inductive systematization we find ourselves involved in a morass of difficulties, most of which are tied to the problem of induction, the problem of justifying the conclusions of inductive arguments. The issues here are many and quite complex. Hempel insists on retaining inductive systematization – the systematizing of inductive generalizations within his characterization of a theory – and Sellars objects to this. Hempel's commitment here stems from his model of explanation requiring laws.

General laws have the function of establishing systematic connections among empirical facts in such a way that with their help some empirical occurrences may be inferred, by way of explanation . . . from other such occurrences. ([16], p. 77)

Both inductive and deductive systematization are concerned with the deriving of empirical generalizations to be used in explanation. Theories are invoked to explain these laws.

Thus we have seen that Hempel believes the role of theories is to explain empirical generalizations. Sellars argues that viewing theories in this way precludes the possibility of fully explaining inductively based generalizations. On the other hand, if theories are construed as concerned to directly explain particular matters of fact, then by telling what those items are and fully accounting for their behavior, we can explain the statistical behavior of those objects over a long period of time or in complex situations. That is, if we know what the object is, then we will know how it will behave. Sometimes it will behave this way and other times in another way. This, according to Sellars, constitutes the proper way to explain generalizations.

Sellars therefore argues against the levels view of theories. He says theories are not designed to explain generalizations, hence there is no need to insist that theories provide for systematizing inductive generalizations. But two problems then emerge for Sellars. First, doesn't the elimination of inductive systematization from the role of a theory throw us back into the theoretician's dilemma? The answer here is 'no'. We find ourselves in the dilemma only if the language in which the observables are described is fully independent of the theory. If, on the other hand, the description of the entities depends on the vocabulary of the theory then theories are necessary and there is no longer any problem.

The second problem is not so readily handled. How does Sellars accommodate all those empirical generalizations? Since scientific theories are not concerned to explain them directly, but only indirectly by telling us the exact nature of the objects they concern, what is their role? Sellars believes empirical generalizations do have a major role to play in the development and acceptance of theories as explanatory frameworks. They are the basis for formulating rules of inference which we organize into a coherent theory.

Sellars believes that induction marks the method of science rather than one minor product of scientific inquiry. We have also noted that Sellars believes that the development of science proceeds within an ampliative framework, i.e., one in which the making of inductive inferences is acceptable. We explain phenomena by replacing descriptions of them with more adequate accounts. Moreover, each successive theory, while more complete than its predecessors if it answers more questions, remains inadequate up to the point where we have a final science. Thus SE requires a framework in which it is rational to move from one set of data to another, both of which are incomplete. Moreover, SE requires that we finally reach a completed science since explanation entails giving a complete account of the objects being explained. And, according to Sellars this can take place only within a completed science.

But this presents *Sellars* with a paradox. For in the same sense that Sellars argued, against Feyerabend, that the framework of common sense is not a theory, we can now argue that Sellars' final theory, FT, cannot be a theory either. If it is indeed FT then it cannot be an explanatory framework. If it is not an explanatory framework it cannot be a theory.

Since common sense has no external subject matter and, therefore, cannot be a theory, FT likewise will have no external subject matter. In the same way that Sellars argues against Feyerabend that to say the subject matter of common sense is common sense objects trivializes the issue, we suggest that to say the subject matter of FT is the universe also trivializes the issue. Just as Sellars argued that theories explain by being 'about' their subject matter and by describing it completely, and that common sense was not 'about' anything, so the final theory cannot explain the universe since it is not 'about' the universe.

FT will provide the framework for constructing accurate pictures of the world; which pictures are matters of fact. Picturing is a linguistic activity. The pictures which represent the world are located in a linguistic framework. Even though picturing is a relation, the peculiar sense in which the relation in question relates anything to anything else gives reason for pause. According to Sellars, a linguisitic picture is a matter of fact proposition. We measure the success of the final theory in terms of the quantity, quality and coherence of the pictures it generates, i.e., the matters of fact it can handle.

Sellars admits the possibility of competing frameworks generating competing sets of pictures. But, he argues, since we cannot answer the question of which framework portrays the *real* world the final decision on which framework to accept will be made on the grounds of the best, most coherent set of pictures which can be produced.

Thus, FT is not 'about' the pictures it generates. FT is nothing more *than* those pictures and the rules for generating them. The pictures then constitute the world in linguistic form. This sounds very much like Sellars' description of common sense as a set of beliefs, opinions, etc. Now if so, the sense in which FT is about something or other vanishes. And if that vanishes, the final theory cannot be an explanatory theory.

If FT cannot in principle offer an explanation, then the very sense in which it is a theory also seems to vanish since, for Sellars $T = F_e$. All of which begins to sound like a *reductio ad absurdum* on SE. For here we have a theory of explanation which requires a theory for explanations in order to be possible in the first case, but which develops the final explanation be eliminating theory in general and the explanation along with it.

Once again we seem to have a case of misleading terminology on our hands. Properly viewed, SE is *not* a theory of explanation. That is, SE *cannot* be a theory of *explanation* since its final product cannot be an explanation. It should be obvious by now that SE does, however, constitute the core of a theory about the development of science. It provides the mechanism whereby we replace theories. Science progresses by replacing old theories with new and more adequate ones. C-rules are used to make the necessary connections.

The question to which we must address ourselves now concerns the framework in which science develops. For it is clear that most of the plausibility of Sellars' case requires that this notion be made intelligible.

CHAPTER V

CONCEPTUAL CHANGE

1. INTRODUCTION

There are two dimensions to Sellars' theory of justification. First, there is the vindication of the inductive policies, the subject of Chapter III. While vindication is a weak method of justification, it is supplemented by the explanationist principle of justification: a theory is justifiable if it provides explanations, Chapter IV.

Much of what Sellars has to say here is influenced by his efforts to separate the process of scientific inquiry from the product of science; thus we *vindicate policies* used in the process of developing theories by producing theories. But given the theory, what justifies our assumption that it ought to be accepted? We justify our adoption of a theory, and hence our belief in what it says, by showing that the theory explains what we would otherwise believe.

Explanationism claims that a theory is justified if explained or if it explains. As it stands this not only sounds reasonable, but worth developing. Explanation is epistemic in nature and the appeal of explanationism lies in the doors it opens. We have been stymied in our attempts to justify theories by appeal first to syntactic and then semantic principles of verification and then confirmation. Now, by enlarging our context, following Goodman's injunction to use everything we can, we may be able to finally break some new ground. That is, it is time to stop approaching justification in a piecemeal fashion and reconsider its general philosophical character.

On the other hand, it is not clear that distinguishing between policies (process) and theories (product) really does leave us in any better position. These doubts arise because the distinction between justifying theories and vindicating policies is not as firm as we would have it if we reject the idea of *FT*, a final science. For we do not vindicate policies which we have used to construct just any theory; rather, we only vindicate policies for a theory which can function as an explanatory framework. Therefore, if a theory is adopted at time t_1 and rejected at time t_2 because it has been replaced by a superior explanatory device, do we thereby rescind our vindication of our inductive policies? Sellars seems to suggest that the final sense of 'vindicate' applies legitimately only in the case of *FT*. Without *FT* we are forced to talk about the vindication of policies with respect to particular theories, and the claim of extreme contextualism again appears warranted. We therefore are forced to *FT*.

Can we make a case for *FT*? I think not. The purpose of *FT*, a completed science, is ill-conceived. The scientific framework cannot rationally replace the manifest. If this conclusion is correct then salvaging the theory of evidence discussed in Chapter IV will require introducing some changes. In the course of modifying Sellars' account here I intend only to sketch an alternative account, one in which the epistemic autonomy of science is retained, but its control over what we see kept in check by the principles governing the framework of common sense.

2. THE SCIENTIFIC IMAGE: A RECONSIDERATION

In 'Scientific Realism or Irenic Instrumentalism' Sellars distinguishes between (1) his *unwillingness* to agree with Feyerabend that common sense is a theory, and (2) the position that he *shares* with Feyerabend, that the framework of common sense, F_{cs}, is false. According to Sellars, while seeking explanations of objects and events encountered in F_{cs}, we develop theories positing theoretical (unobservable) entities. Theories are *about*, initially, events which are described in the language of F_{cs}. F_{cs}, however, is *not about* anything in that same fashion, *ergo* not a theory. In the long run, by claiming that "in principle it is the framework of theory *rather than* the observation framework which is real", he intends to explain how F_{cs} can be false.

Unfortunately, Sellars does not clearly articulate the important differences here between claiming F_{cs} is *false*, and F_t is *real*. The reasons for the latter are tied to Sellars' teleologial view of science. He claims the falsity of F_{cs} on somewhat different grounds.

Let me temporarily leave the question of content out of the discussion and concentrate first on issues of logical priority. A precise specification of the meaning of 'framework' is quite difficult. Consider, however, the following as a working characterization. Frameworks are the contexts in which the general pictures Sellars calls 'images' are generated. Thus, F_{cs} is the context in which we formulate the manifest image of man in the universe. But how are we to understand what the manifest image is? Is it the common sense view of the world as it looks at the time when we have a final science which can be presented as a real rival?

If we are to consider the manifest image in terms of the picture it generates today, then it would seem that there is no arguing the truth or falsity of either framework. On the one hand, there is no good reason to assume that the present scientific image is true, especially on Sellar's own grounds. The present state of science is one of becoming. On the other hand, it is unfair and patently unreasonable to assume that the finished *future* science is to be compared with the *present* manifest image. For why should the manifest image not develop even further, or along with the scientific? If it does, then it should be compared with the scientific image at the last possible stage, anything else would be arbitrary.

Talk about comparing future and present images is all well and good, but does little for a firm understanding of the real problems at issue here. Basically the matter comes to this: why should we, at whatever future point, reject the manifest image? If the answer is because it is false, we still need to know why we should reject false images. Another way of putting this is to ask in which context do we locate the principle that authorizes the abandonment of one framework in favor of another? It *should* be located in the framework of common sense. For it is in that context that science was licensed as an activity in the first place. Loosely speaking, we should take common sense's notorious hardheadedness into consideration at this point.

If we ask what *really* initiated the refinements of the 'original' image which led eventually to the refinements of science, we would truly be hard pressed for an answer. But on any account it would be obvious that attempts to refine and make more precise the framework in which we think about the world are motivated primarily by the need to act on whatever knowledge such refinements can produce. And the sense in which we produce 'refinements' of the framework we are using is the sense in which we render that framework more responsive to our needs. Baptize this principle of evaluating the degree of refinement in terms of the consequences of action: *PPI*, pragmatic principle *I*.

Referring to *PPI*, we can understand postulating imperceptibles in the first place and, hence, the beginning of theoretical science. Science did begin because of the need to explain, as Sellars suggests. But the need to explain is not an end in itself. It is funded by *PPI* to seek explanations to ensure the success of our projects.

If *PPI* funds scientific inquiry, then it is but a short step to the next point. Common sense is renowned for its insistence on no nonsense. And if the framework of common sense does produce the manifest image, then it is the same framework which initiates science. But science is, in this sense, a part of the framework of common sense. For it would be most contrary to common sense to initiate and encourage the expenditure of all the energy that goes into scientific inquiry without expecting to get something out of it in the long run. Thus, it would be reasonable to assume that there is no conflict between science and common sense in the long run. For if *PPI* is indeed operative, then common sense would simply adopt whatever science has to say once it is convinced that science can indeed answer the questions. This gives us *PPII*: accept results of inquiry which bear on problems to be solved, *ceteris paribus*.

This suggestion entails that common sense itself be constantly undergoing the same sort of changes that mark a developing science. Among other things, common sense contains both beliefs and prototheories of its own. We know that beliefs change and that prototheories are either abandoned or developed. So we can say that common sense changed without running into too much trouble. And, if common sense changes, then the image of the world which it projects must also change. And to the extent that common sense is convinced of the value of the product of scientific inquiry, it will come to adopt the fruits of science. The acceptance of science is, therefore, a possible common sense activity. In other words, the acceptance of the scientific image of the world is not only a reasonable but a very possible function of common sense. And even if such acceptance entails the further refinement of the conceptual basis of common sense, CB_{cs}, this is easily accommodated by adopting a principle compatible with both *PPI* and *PPII*, *PPIII*: work out consequences of adopting new data, sorting out inconsistencies.

Sellars argues for replacement of CB_{cs} by *FT* on the grounds that *FT* and common sense will be at odds. On the other hand, I am arguing there can be no *FT* because the purpose for which it is to be developed is ill-conceived. The alleged purpose of *FT* is to replace common sense. The scientific image is generated by *FT*, and the manifest image by common sense. *FT* then is to replace common sense. It is to replace common sense by explaining it. This is to be done by redefining the terms and replacing the rules of thought embedded in common sense with theoretical correlates.

But, if common sense, or the manifest image is not in conflict with science, as I have argued, then there is a very good sense in which *FT* is unattainable. For if the acceptance of the scientific product is a function of common sense principles and, furthermore, if this acceptance is crucial to our understanding of the role of science, then it is inappropriate to speak of the scientific image replacing the manifest. The activity of science, the postulating of imperceptibles to explain perceptibles, is funded or licensed by common sense on the basis of our need to know in order to act. That point is crucial, since even if *FT* were possible, for it to replace common sense requires that it be accepted or adopted. But what could possibly be the reason for adopting science? To say that we adopt *FT* because it explains is to give only a sufficient, but not a necessary condition. Knowledge is for the sake of action. But for *FT* to be adopted in this way, for the sake of action, is for *FT* to become part of common sense, not to replace it. It results in the scientific image becoming the manifest image. It might be argued, however, that these are mere quibbles. What difference does it make if science replaces common sense or vice versa? The result is the same; science wins out.

But the question here is not whether science will win, but whether *FT* is possible. And the *FT* at issue is the one characterized by Sellars as designed to replace common sense. But if common sense is presupposed, that is, if the success of our efforts to build *FT* depends on the possibility of *FT* being *rationally adopted*, then the sense in which *FT* replaces common sense by explaining it is lost. For if common sense is the active ingredient here, it is not explained at all, it *uses* the picture generated by science.

Another way to approach this is to claim that from the start Sellars was misguided in his efforts to identify something as 'the manifest image'. For the manifest image is a function of common sense, and common sense is not a static body of beliefs and principles, but rather a constantly changing body of beliefs and principles, values, hunches, outmoded ideas and visionary projections. In fact, if forced to characterize common sense in terms of something more substantial than the previous list of changing objects, I would suggest that its chief characteristic is just that set of principles which encourages new ideas, helps digest unsystematized data and establishes priorities for action. And if this characterization is anywhere near the truth, then we can begin to appreciate more fully the sense of Sellars' claim that the basic principles of common sense have a methodological hold on us. What he failed to realize, however, was exactly how fundamental a hold this is.

Part of the problem of understanding Sellars here is determining the extent to which he is committed to the idea that *FT* is an actual possibility. And while he does sometimes sound as if he merely means *FT* to be kept in mind as a regulative ideal, he also seems to believe that *FT* is actually constructible. Thus, on the one hand, he says,

the perspective of the philosopher cannot be limited to that which is methodologically wise for developing science. He must also attempt to envisage the world as pictured from that point of view – one hesitates to call it Commpleted Science – which is the regulative ideal of the scientific enterprise.

On the other hand, in talking about the methodological hold the principles of common sense have on us, he claims that,

It is the rock bottom concepts and principles of common sense which are binding until a total structure which can do the job better is actually at hand – rather than a 'regulative ideal'.

But clearly, he must give up the idea that *FT* can be developed in line with the goal he has in mind. This then leaves us to consider the sense in which *FT* can function as a regulative ideal. It would seem however, that even as a regulative ideal the concept of *FT* must be abandoned. For there is simply no justification available to warrant it. Adherence to a wrong-headed belief can lead us astray. In this case, clinging to the ideal that science should be conceived as progressing toward some unified completed stage in which it takes over from common sense simply does not make any sense. It particularly fails to impress when you consider the consequences of Sellars' program. The key point, following from his commitment to scientific realism, is that the perceptible objects and processes of everyday life are not real. The final determination of what is real is science's job. One of the problems to be overcome before we can abandon the manifest image is to manage an explanation of what it is that we see in terms of the ultimate constituents decided upon by science. But that presents no real problem on Sellars' view, for we merely have to engage in a series of redefinitions using correspondence rules, a variety of which is already operative when, for example, we say that the table is really a collection of electrons, etc.

Now I certainly do not wish to argue that it is electrons which aren't real. Nor do I wish to play the modern Dr. Johnson and break my toe in order to prove to Sellars that rocks really are real. For that surely is not the point; Sellars does not conceive of himself as the modern Berkeley. The heart of the matter lies in the manner in which Sellars himself approaches the issue. The original image, of which the manifest image is a refinement, was an image of a world composed only persons. All the objects and processes of nature were endowed with human attributes. Thus, the wind was angry, the ocean furious, old trees wise, etc. The manifest image is a result of refinements by which the objects of the original image are being depersonalized so that as persons we see a world of objects. Persons and objects form the two primary logical subjects in the framework of common sense. So far, the primary logical subjects of the framework of science have not been established, but a good guess is that one of them, if there is to be more than one, will be number. Behind the urge to replace the manifest by the scientific image is a principle which argues against ontological crowding, hence the elimination of sets of logical subjects in favor of another. Let us call this principle the principle of ontological simplicity, *OS*. And clearly, if we are to challenge Sellars without reducing ourselves to mere instrumentalists we must challenge *OS*.

3. ONTOLOGICAL NECESSITY

To counter *OS* in a manner which speaks to Sellars' discussion and which also provides a way out of a rather unhappy situation, we must first get to the bottom of Sellars' attack on the manifest image. There is, to begin with, his association of the manifest image with what he calls the perennial philosophy of man in the world, another construct introduced for purposes of polarizing his discussion.

This construct, which is the 'ideal type' around which philosophies in what might be called, in a suitably broad sense, the Platonic tradition cluster, is simply the manifest image endorsed as real, and its outline taken to be the largescale map of reality to which science brings a needle-point of detail and elaborate technique of map-reading. ([45], p. 8)

While it may be the case that instrumentalists have accepted the manifest image as real and view science as only a useful tool, it does not follow that that is the only way in which to deal with the manifest image. Historical issues and philosophical battles aside, that there is something like the manifest image of man in a world I take to be correct. But to deny that tables and chairs, trees and flowers, seas and mountains are real because a certain misguided philosophical school likes them is going a bit far. Granted this is not Sellars' reason for suggesting that we replace the manifest image. But it does appear that Sellars is throwing out the proverbial baby with the bath water. His attacks on instrumentalism stand independent of his concern with the manifest image.

So let us assume that we are here concerned only with the manifest image as that picture of the universe we live in shorn of scientific results and superstition. This is the world a philosophical naturalist like Quine is constantly citing as a starting place and constantly trying to escape, the world we live in as biological beings. But to assert that because of certain biological characteristics which permit perception of objects of certain sizes, sounds within certain frequency, sizes and colors falling within only part of the spectrum, that those features hold absolute primacy is, as both Quine and

Sellars note, to deny free range to our other faculties. But just as it would be absurd to hold that the world of objects and processes falling within our perceptual range is all that is real, like claims for the world of imperceptible entities or the world of macro-entities, such as galaxies, are unreasonable. If we turn our attention to one particular object which falls within the scope of things which exist in a world of the manifest image, paintings, we are going to have a very difficult time explaining that what we see isn't really a painting, but imperceptible entities of a certain kind organized in a certain way. Now it might be objected that paintings are hardly the kind of object with which Sellars is concerned. But I can see no way of ruling them out except in an extremely arbitrary way.

Objects like paintings play an important role here because they are not only conventional devices, but nonreducible. And, by way of extension, I would argue that a tree is also nonreducible. This is not to say that the behavior of a tree cannot be explained in terms of scientific laws. But the tree can be an object of our attention despite what we say about the ultimate cause of its behavior. And there I think we have the crux of the matter. A large class of items intrude upon us in a manner best characterized as perceptual despite what else might be true of them. This is also independent of the role of language, which is not to deny that language has an important role in helping us organize our response to the world. And while it might also be true that depending on the presuppositions embedded in language about the kinds of things there are and the types of inferences permissible, which information can be greatly influenced by the results of scientific research, it is also the case that these factors are independent of our hearing range, visual range, etc.

The conclusion to be drawn from these remarks is that given certain facts about ourselves, and that we live in a world which initially, in an unreflective way, is quite different from the world pictured by science, but quite real, that arguments for *OS*, however refined, tend to over-simplify a very complicated situation. But if we reflect that the principles which license science are principles of the framework of common sense, then perhaps with a little more reflection on the characteristics of common sense we might make some headway.

To begin with, it seems to be a piece of common sense, that one person, e.g., Jones, can play two roles, e.g., philosopher and farmer. Furthermore, that a person can be regarded in this way does not thereby create two people and contribute to the problem of overpopulation. That is, it seems possible to accept such a situation without entertaining any doubts about the structure of the universe. Furthermore it seems quite possible to state without fear of confusion that it is the same person, i.e., that the notion of a person here is not exhausted by or reducible to its roles. Thus it would seem that another principle of common sense would be that the same object can be viewed in a number of ways. Not only can it be viewed in a number of ways, but of each role we might use a different language. Furthermore, the languages might be incommensurable in the sense that there is no counterpart of 'logician' in farmerese. But despite certain aspects of incommensurability like this, these roles are common to an entity designated by the same name. The name serves as a common touchstone into which each language ties. Moreover, to object that we must have two people because despite sameness of name there is no point at which both languages are appropriate is false. For it seems perfectly possible, even likely that while plowing a field a farmer is simultaneously doing philosophy. And finally, the choice of language you use to talk about Jones will be a function of what it is you wish to say about him. Thus while knowing that Jones is a philosopher you never talk about him in the appropriate way because your only concern is with Jones, the farmer.

In this way I think we can see our way into the problem of ontological primacy and out again. In first approximation, we can say with Carnap that what is real is a function of the framework within which you are operating. That it is real will be a conclusion based on a necessity characteristic of the framework. The necessity characteristic of a framework is that property which, if denied or ignored, renders that way of looking at things useless. Thus, physical objects are real in the context of common sense because to deny this would invite physical disaster and the purpose of the framework is survival. Imperceptibles are real in the context of science because they are what explains and the purpose of science is to explain. If in order to survive we need to know why it is that object x behaves in manner y, we turn to science for the answer. To conclude that what is real is a function of which is necessary seems not only reasonable but obvious.

4. REASONABLENESS AND RATIONALITY

There is no reason for conflict between common sense and so-called scientific ways of looking at the world. One thing may be reasonably looked at in more than one way without having to reduce one to the other. If I am right this would undermine the assumption of the ontological primacy of theoretical entities. It also sets the stage for undercutting the claim that the scientific view ought to be preferred. It is one thing to argue that the results of the scientific inquiry ought to be adopted so as to enrich our understanding and quite another to argue that understanding results only from adopting a scientific point of view.

What is ontologically necessary are those objects the denial of which would render an entire way of speaking empty. Sellars suggests that once we have a fully articulated science the need to rely on the imprecise language of common sense will disappear. But the point is that while the language of common sense may be imprecise it is only imprecise for the purposes of science. The degree of precision necessary is a function of what is at stake. For the purposes of survival, a language which directs your attention to objects within your perceptual field may be sufficient. In fact, a language in which there are no words for trees may be counter-productive, too much precision may be harmful. The failure to recognize the full consequences of the view that different dimensions of human action have different requirements lies behind the claim that someone who fails to accept the scientific point of view with regards to a given object is irrational.

I have, in effect, been arguing that there is no preferred conceptual framework. The choice of framework is a function of what is to be accomplished. Furthermore, there is nothing wrong with the idea that one person may use many frameworks in the course of working his way through the day. But I have also suggested that even though accepting a framework involves accepting its entities as real, this does not entail ontological crowding. The example of the farmer may not be strong enough since, as a matter of fact, I do have recourse to a meta-language in which I can talk about the person who plays both roles. But if we shift to the old example of the table/collection of molecules, where the explanation of certain features of the table is in terms of its molecular make-up, the question concerning the number of objects is harder to answer. Is there a table, a collection of molecules, or both? I want to claim there are two things, the table and the collection of molecules. But from this it doesn't follow that these things exist in the same context. Moreover, it does not appear to be a reasonable worry.

To ask what a particular item *is*, is to do one of two things, either (a) to request information concerning the name of the object, an indication of the type of thing it is supposed to be and some clue as to the proper way to use the terms associated with it, or (b) to ask for an explanation of the thing, which is to seek an account of why it does what it does. Thus, to ask what a table is, is to ask what that thing there is called and what it is supposed to do and what sorts of properties it is supposed to have, since I would like to know how to use the word 'table'. To go 'further' and ask what a table *really* is, then is to either engage in the noncognitive search for essences or to request an account of why the thing in question manifests the properties it does. This latter always involves appeal to another framework. In this case the framework of physics serves. Crowding is avoided because ontology is a function of framework and there are no tables in the framework of physics.

What the thing is then remains a legitimate question only in the context of asking for an explanation of some of its properties, which explanation cannot be provided within the framework of common sense. At that juncture we initiate our own version of the translation game. We seek a clear correspondence between the language of common sense and the language of the explanatory framework we choose. Once again, how much clarity will be a function of what the objective is. But to ask for undeniably precise correlations is to step outside the bounds of reasonableness, for the correlation itself is not the end. It is only a step in a more general exercise and there is a practical limit which will do. The kinds of reasons one adduces for stopping will be practical reasons concerning the relation between the benefit to be gained from a further search for precision and the value of the explanation offered at this level of clarity.

The moral to be drawn from the above account is that in offering explanations for the behavior of an object by appealing to objects postulated in another framework, we are not explaining what the thing is, but why it behaves as it does. The question of what an object *is* is unintelligible outside the framework in which the terms concerning the object have an established usage. A corollary is that explanation is a practical as opposed to a purely epistemic notion.

Hopefully, this approach provides a way out of the problem of having to decide between common sense and science as to the ultimate make-up of the world. The decision is already made when you choose the activity you are to perform. You live in a world of cars and highways when attempting to make your way to the lab. But once in the lab the presuppositions of current theory in high-energy physics take over. Now this need not lead to a kind of ontological schizophrenia, as might be feared. To recognize that you work with different types of things depending on what you are doing is to recognize the complexity of the world whenever a sentient biologically limited being plays the role of interpreter. To seek unguarded simplicity is to look for a world in which man does not function.

5. CONCEPTUAL CHANGE

If I am correct and there is not only no conflict between the manifest and the scientific image, but also that there is no real worry over ontological crowding, then we can now turn our attention to the problem of conceptual change.

The basic scheme within which I wish to work is the following. Consider *CF* our conceptual framework. At the uppermost level it consists of *PPI, II* and *III* and CB_{cs}, the categorial basis of common sense, containing generic rules of inference, transformational schemata, fundamental categories, fundamental formation rules — all the sorts of things constitutive of a conceptual scheme. One of the more important formation rules will be the one which permits the establishing of new categories in response to demands generated by activities using other aspects of the scheme. Within the scope of CB_{cs} fall a number of conceptual schemes, viz.: science, S_s, ethics, S_e, etc., each tied to a particular aspect of being human, e.g., explanation of physical events, interpersonal relations. (There are a number of other conceptual schemes which form part of our conceptual framework as well. For example, there is one for art, one for social behavior.) Thus, we can talk about science as part of our general conceptual framework without worrying about science as an inhuman way of seeing the world, etc.

So much for vague preliminaries, what about change? First we need to distinguish various aspects of change. For not only can S_e or S_s change, they can change with respect to both form and content; likewise for CB_{cs}. Furthermore, it is not clear that changes in, for example S_s necessitate changes completely up the ladder. In other words, a slight change in a part of one scheme does not mean we have a new conceptual framework on our hands. Nor is it the case that even wholesale changes in one scheme entail consequences for another one. Thus, while astronomical theories come and go placing man here and there in the context of the universe, it still remains true that if we don't want to hurt ourselves we had better avoid walking into a tree.

On the other hand, this does not mean that questions arising out of activities conducted in the context of one scheme may not have bearing on other schemes. Thus, as we have seen, science is a direct result of the desire to know why trees hurt you when you bump into them. Furthermore, communication between schemes is not only important but a fact concerning the way we live. If we want to know what causes the strange behavior for which there is no apparent cause, but for which the conceptual scheme of mysticism has an explanation, we can appeal there, or ask science. Whichever explanation suits us we adopt as part of our understanding. If enough people adopt it, it becomes part of the content of common sense.

There is also the question of genesis. I would like to suggest that the conceptual scheme of objects and visual processes has a kind of priority because it is most closely associated with the initial problem of survival. Science and ethics are outgrowths from here which have taken on independent status. That they too bear on survival is also true. But it is not basic gut survival, rather survival in terms of new goals and visions of what is possible. The process is one which can be characterized by analogy to inventing languages to discuss new aspects of things with which you are already familiar. Furthermore, I think this is a very strong analogy, for knowing your way around S_s requires knowing the language of science, where knowing a language here involves all the considerations brought out in Chapter II. But, whatever the facts of genesis are, they are now unimportant. What remains is that our various conceptual schemes can be viewed as different languages 'about' fundamentally the same world, where use of various languages commits you to levels of reality, where levels here need not necessarily entail horizontal layering.

Let us turn now to questions of change in science and see how to apply the results of this investigation. To talk to change in science we need to distinguish two sorts of changes: changes in methodological rules, and changes in theories. To the extent that we change the methodological rules we change the constitutive nature of science. These rules fall into three categories: rules for obtaining what we might call scientific claims, rules for testing scientific claims and rules for accepting and rejecting claims. Under the first heading we find rules licensing both induction and the deduction of claims from theoretical assumptions. Under the second category falls the general area of confirmation theory. It is the third category which concerns us most now.

The most important dimension in accepting and rejecting claims generated and tested in a scientific context is that of the role and status of evidence. As we have seen, Sellars' theory of evidnece has some problems which are hard to overcome. Nevertheless, Sellars' major point is well taken: different types of claims have different kinds of considerations governing their acceptance. Quine and Ullian make a similar point in *The Web of Belief*. Reasons for accepting a statement as worthy of use and further investigation vary, depending on the kind of statement. When Sellars says that the meaning of 'probable' is 'there are good reasons to accept' and, furthermore, that there are different senses of 'probable' depending on the kind of statement with which we are concerned, he is after the idea that despite the fact that all sentences require some reason to be accepted, the type of reason will vary with the type of statement. His appeal to practical reason as the heart of probability reasoning simply amplifies the idea that it is the giving of reasons that is important. When he ran into difficulty with first order probability statements he did so because those statements did not provide us with the means for completing the practical argument resulting in the conclusion 'I shall accept that p'. What was missing was the wherewithall to cite evidence and the logical connection making possible the deduction of the conclusion. As an alternative I want to suggest first that what pass for first order probability statements on Sellars' account are statements like observation sentences and that we modify his account by admitting observation sentences if they are semantically assertable. Furthermore, let us abandon talk of probability here, for what is really at stake is the acceptability of statements, not their probability. Thus, we might extend the discussion by claiming that for any sentence S it is acceptable if semantically assertable. Now to say S is acceptable does not entail that S is true. Rather it means that S is worthy of inclusion in the body of statements comprising the theory with which we are working. To include a sentence within a theory is to acknowledge that it is worthy of further analysis and testing. All of which says nothing about the ultimate value of S or of the theory.

It may be objected at this point that Sellars uses the concept of semantic assertability in his account of truth and that we have hereby confused not only truth and evidence, but evidence and reasons. Two points will serve to clear this matter up. First, a theory of reasons is not a theory of evidence *simpliciter*. What counts as evidence is one thing. It involves criteria for identification of X as a possible evidential claim. The giving of reasons involves not only knowing what can be cited as evidence, but under what conditions it can be so used and these are often goal dependent. Secondly, Sellars' theory of truth, while indeed phrased in terms of semantic assertability also entails reference to a set of propositions at some future point which accurately picture the world. Thus, in *Science and Metaphysics*, he says that while S may be S-assertable at time t_1, it is true if and only if included in the set of propositions asserted as true at t_2. Thus, we can modify his original claim that S is true if semantically assertable to S is believable as true if semantically assertable at t_1. And this hardly conflicts with our intuitions if we remember that what we cite as evidence is a statement we *believe* to be true. If it eventually turns out not to be true for various reasons such as following from a theory we reject, we rarely find ourselves in a bind.

It seems quite reasonable to accept Sellars account truth as true-in-the-long-run. It does not commit us to anything like a final theory in the sense in which the scientific image did. Furthermore, the idea that for S to be true

it must remain true in whatever theory we accept is both reasonable and
necessary if 'true' is to carry any weight at all.

When we turn to other kinds of statements involved in making up a theory
we need similar rules for accepting them. Sellars' account of some of the
considerations involved in accepting rules of inference seems well presented.
We do accept those statements from which only true sentences are inferred
and we reject those which fail on this account.

Change enters the picture when for one of two reasons things don't work
out as they are expected to. Borrowing familiar terminology, there are internal
and external reasons for initiating changes. The internal reasons are internal
with respect to the conceptual scheme of science. They are not internal to
the theory. Rules governing acceptability of claims generated in the context
of using a theory are formulated in terms of methodological rules and com-
pose one of the two parts of a conceptual scheme already outlined. They may
also involve appeal to a third aspect of a conceptual scheme, that is the ideas
with respect to which the scheme is developed. Thus, if we find a rule which
says something to the effect that inconsistencies in experimental results are
to be worked out in accordance with the maxim of trying to save the theory,
we find here an implicit appeal to consistency as a virtue to be maintained.
This then gives us another level to consider when we speak of change. It is
also quite possible that the 'ideals' merit revamping for one of two reasons,
either by following them we fail to meet the objectives set out for science
by common sense or because of violating some ideal in another scheme,
which, for practical reasons, bring pressure to bear on the techniques of
science, viz., the problems of elimating religious bias from investigations of
the heliocentric view of the solar system.

Not only can the ideal be changed, but the methodological rules themselves can be changed. This can result from noticing that one of the rules
is inconsistent with the others and, thereby, produces results incompatible
with the ideals. It can also be due to a change in the ideals. Consistently
inconsistent results in one part of a theory that has otherwise stood quite
well might also warrant re-evaluation of the methodological rules.

Accepting or rejecting rules then is a function of internal considerations
of the scientific ideals, rules, and theories involved. It is also a function of
external considerations such as whether or not the theory produced by using
the rules satisfies the need for which it was developed. If, for example, we
produce a theory which fails to explain some phenomenon to the satisfaction
of those who seek the explanation, then it seems quite possible that the rules
under which the theory was developed would have to be revised. But, it

would be objected, that would be irrational. If the theory meets all the requirements for a good theory, then it should be accepted. But on what grounds? If an argument showing the necessity of the acceptance could be arranged it might work. But notice this would make scientific considerations prevail over those of common sense, which would be to play the game backwards. Ultimately the acceptability of science is a function of whether or not its results fit a need. And if they don't, there is no way the argument could be made to go. Now it might possible to argue, in common sense terms, that the need is ill-conceived. That is, it might be possible to show that while this theory doesn't answer the specific need in question, by accepting its results that need may be seen to be illusory in the first place. But this argument must take place in the context of the principles and ideals of common sense. It is not a scientific argument, but a practical one. If successful it will perhaps have repercussions for the structure of CB_{cs} and this may in turn generate alternative changes in S_s.

What I have tried to demonstrate is that once the structure of a conceptual scheme is laid out, the variety of ways in which it might change, the variety of reasons for conceptual change, and the possibilities for success are extraordinarily intertwined and complex. Furthermore, we might want to distinguish between changes in a conceptual scheme and conceptual change, where the latter entails the abandonment of an entire framework such as *CF*. But to do this entails being able to make a case out for alternative frameworks. I think this is impossible, if they are frameworks for us. For, if I am correct, the framework of common sense is what generates the manifest image of man in the world, man as a biological being with certain inherent abilities and limitations. That image of the world is common to all men. How it is embellished and interpreted is how it becomes a world in which *we* live, since the embellishments concern how we explain this world, evaluate it, and claim we ought to behave with respect to it. And in the process of elaborating *CF* by adding conceptual schemes of various complexity and relatedness we indicate the difference between mere men in the world and being human.

6. RATIONALITY VERSUS REASONABLENESS

In arguing for changes, we constantly face the task of justifying our demands. To be able to do so in such a way as to satisfy all reflective men would be the hallmark of rationality. But given the complexity of the issues involved in conceptual change we should reevaluate our concern with rationality. In those cases where changes are not merely acquired, but where there is a case of decision and acceptance, it is one of community decision. I can conceive of no situation in a society where one man's decision singlehandedly produces a conceptual change of any degree. To change the conceptual structure involves having suggested changes accepted by many. And the reasons for acceptance will vary from individual to individual. In fact we can go so far as to claim that each individual might choose quite rationally to adopt the changes, but that we could still not claim it was a rational change. For rationality in choice is a function of an individual maximizing his values and goals with respect to the available information. It is quite possible to imagine a situation where most of the goals of the choosers vary, as do most of their values, but where the same decision is reached, for different reasons, if you will. Under these situations to call the change as it affects the entire community a rational one hardly follows. For rationality is not a property of communities, but rather of individuals.

On the other hand, it would not be strange to claim that the change was a reasonable one provided it could be explained by appeal to social and/or epistemic considerations. Finally, arguing for reasonability of change over rationality seems preferable when we consider that the reasons for instituting changes are a function of changing goals and values. To speak of rational change where conceptual schemes are concerned is to invoke the spectre of some absolute standard or final goal in terms of which to evaluate alternatives. But ultimate goals and values change. Moreover, they change in response to activities undertaken to implement them, and often before the goal is attained as well. In other words, a very complicated feedback mechanism is involved. Given certain goals we attack certain problems. Solutions to the problems help shed light on the nature of the goal, or put the goal in a new light which causes a re-evaluation of its place as a goal, and so on. By making certain objectives possible, such as total participatory democracy, technological advances initiated for a variety of reasons give cause to reconsider whether or not we really want to involve everyone in all government decisions, even though by means of television and complicated telephone hookups we could manage it.

To argue for the rationality of change in the face of a situation which is itself characterized by constant change is unreasonable. But to characterize change as reasonable seems plausible if we can meet one further point. One of the hallmarks of reasonableness is explanatory power. Thus to speak of the reasonableness of change entails being able to explain change. And it does not detract from the enormity of the development if the best we can do is retrospectively, and maybe even, contemporaneously, explain the changes that took place and their ramifications. For by denying that scientific change is rational all that is lost is the image of science proceeding toward some ultimate goal which may not be attainable. And surely without good cause to believe in such a goal we would do better to concentrate on understanding what it is that change really entails.

NOTES

CHAPTER I

[1] 'Traditional' here refers to the investigations of logical empiricists dating as far back as William Whewell.

[2] The straightforward claim that the product of scientific inquiry is the only proper object of study is rarely explicitly stated, but see Nagel [24], pp. 13–14 and R. Rudner [35], p. 8. On the whole, the motivation for the construction of formal languages and the attempt to characterize formally such concepts as 'verification' and 'evidence' can best be understood by realizing that, for a large group of philosophers, e.g., logical positivists, what science *actually* is and what scientists do is irrelevant to the specification of the meanings of these concepts. Clarification is only possible, they contend, through the construction of formal languages in which ambiguities can be eliminated.

[3] See also Popper, [29].

[4] Recently there has been a shift toward investigations into the history of science for clues to scientists' methods. By appealing to history, merely abstract philosophical analysis is rejected in favor of a case by case historical analysis, and, subsequently, an effort to schematize the structure of science. See especially Hanson [13], and Humphreys [19], Lakatos [20], and Laudan [21]. P. K. Feyerabend's rejection of the standard analytical approach in favor of the historical provides a pronounced example of a philosopher's dissatisfaction with our general preoccupation with method. The development of Feyerabend's thought (which appears to be in a constant state of evolving toward an ever-receding epistemological extreme) can be traced from a classical analytic-type objection to Hempel's theory of meaning, [5], through a transition stage where he begins to argue the historical case, in both [7] and [9], through two recent papers: [3] and [10] to the apocalyptic [4].

[5] By 'Humean' I mean the traditional raw empiricism which eschews the necessity for postulating theoretical entities and relies only on basic observation.

[6] The allusion here is to the problem of explaining a single case using statistical generalizations; for references see [17] and [37].

[7] Short Note on Notation: When discussing standard positivistic accounts I use 'NL' to abbreviate the phrase 'neutral observation language', as opposed to something more in line with the general notation for L, i.e., L_O, L_{Od}, etc. standing for 'observation language', 'descriptive observations', etc. An alternative to NL which appears to produce a greater degree of uniformity is L_n. But because the neutral observation language represents an ideal case I prefer to emphasize the point notationally. Likewise 'FT' represents an ideal, even though Sellars sometimes treats it as something more substantive. I will continue to stack up capital letters when ideal concepts are at issue.

[8] The essentials of this view are contained in [49].

[9] 'DN' is used here to refer to the deductive-nomological theory of explanation developed by Hempel and Oppenheim in [16].

10 In a sense the point is trivial, Sellars seems to have it by definition: logic − sound; sound − truth-preserving; only deduction is truth-preserving. On the other hand, this account of 'logic' forces attention to the problem of what else discovery could require if not a logic.

CHAPTER II

1 In this light correspondence rules play a role analogous to constitutive rules. See Chapter IV.

2 While agreeing on the general need for criticism, Popper has a slightly different view of its nature and role; See [30].

3 Something like this seems to be behind Laudan's correct rejection of truth as a deciding factor in theory choice in his [21].

4 Since this is not a complete examination of Goodman's epistemology, I feel justified in limiting myself to *Fact, Fiction, and Forecast*. But for those who are interested in Goodman's direct handling of this issue, see [11].

CHAPTER III

1 It is logically prior in order of justification, not in the order of creation. The twist here is that Sellars is talking about these relations as if they were comparable, when, in fact, they belong to different contexts. The notion of logical priority is, in this case, an issue concerning the reconstructed finished theory.

CHAPTER IV

1 For some preliminary efforts along this line see [26] [27].

2 See Sellars' discussion of the distinction between 'all-statements' and 'lawlike' statements in [40], pp. 289−299.

3 The modification consists in shifting the emphasis to the systematizing function of theories.

4 See, however, Sellars' attack on Feyerabend, for his overemphasis on the perceptual role of theoretical concepts in [46], pp. 351−354.

BIBLIOGRAPHY

[1]   Carnap, R.: 'Empiricism, Semantics, and Ontology', in *Meaning and Necessity: A Study in Semantics and Modal Logic* (University of Chicago Press, Chicago, 1947).

[2]   Feigl, H. and Brodbeck, M.: *Readings in the Philosophy of Science* (Appleton-Century-Crofts, New York, 1953).

[3]   Feyerabend, P. K.: 'Against Method: Outline of an Anarchistic Theory of Knowledge', in *Minnesota Studies in Philosophy of Science*, Vol. 4, ed. by M. Radner and S. Winokur (University of Minnesota Press, Minneapolis, 1970).

[4]   Feyerabend, P. K.: *Against Method* (New Left Books, London, 1975).

[5]   Feyerabend, P. K.: 'An Attempt at a Realistic Interpretation of Experience', *Proceedings of the Aristotelian Society*, New Series, 58 (1958).

[6]   Feyerabend, P. K.: 'Classical Empiricism', in *Methodological Heritage of Newton*, ed. by R. E. Butts and J. W. Davis (University of Toronto Press, Toronto, 1970).

[7]   Feyerabend, P. K.: 'Explanation, Reduction and Empiricism', in *Minnesota Studies in Philosophy of Science*, Vol. 3, ed. by H. Feigl and G. Maxwell (University of Minnesota Press, Minneapolis, 1962).

[8]   Feyerabend, P. K.: 'In Defense of Classical Physics', in *Studies in the History and Philosophy of Science* 1 (1970).

[9]   Feyerabend, P. K.: 'Problems of Empiricism', in *Pittsburgh Studies in the Philosophy of Science*, Vol. 2, ed. by R. Colodny (Prentice-Hall, Englewood Cliffs, 1965).

[10] Feyerabend, P. K.: 'Problems of Empiricism, Part II', in *Pittsburgh Studies in the Philosophy of Science*, Vol. 4, ed. by R. Colodny (University of Pittsburgh Press, Pittsburgh, 1970).

[11] Goodman, N.: 'Sense and Certainty', *Philosophical Review* 61 (1952).

[12] Goodman, N.: *Fact, Fiction and Forecast* (Harvard University Press, Cambridge, Mass., 1953).

[13] Hanson, N. R.: *Patterns of Discovery* (Cambridge University Press, New York, 1968).

[14] Hempel, C.: 'Inductive Inconsistencies', in *Aspects of Scientific Explanation and Other Essays in the Philosophy of Science* (Free Press, New York, 1965).

[15] Hempel, C.: *Philosophy of Natural Science* (Prentice-Hall, Englewook Cliffs, 1966).

[16] Hempel, C.: 'The Theoretician's Dilemma', in *Aspects of Scientific Explanation and Other Essays in the Philosophy of Science* (Free Press, New York, 1965).

[17] Hempel, C. and Oppenheim, P.: 'Studies in the Logic of Explanation', *Philosophy of Science* 15 (1948).

[18] Hume, David: *A Treatise of Human Nature* (Clarendon Press, Oxford, 1888).

[19] Humphreys, W. C.: *Anomalies and Scientific Theories* (Freeman, Cooper & Co., San Francisco, 1968).

[20] Lakatos, I.: *The Methodology of Scientific Research Programmes*, Philosophical

156

Papers, Vol. I, ed. by J. Worrall and Gregory Currie (Cambridge University Press, London, 1978).

[21] Laudan, L.: *Progress and Its Problems* (University of California Press, Berkeley, 1977).

[22] Lehrer, Keith: 'Justification, Explanation, and Induction', in *Induction, Acceptance, and Rational Belief*, ed. by M. Swain (D. Reidel, Dordrecht, Holland, 1970).

[23] Kuhn, T.: *The Structure of Scientific Revolution* (University of Chicago Press, Chicago, 1962).

[24] Nagel, E.: *The Structure of Science* (Harcourt, Brace & World, Inc., New York, 1961).

[25] Peirce, Charles Saunders: 'The Fixation of Belief', in *Philosophical Writings of Peirce*, ed. by J. Buchler (Dover, New York, 1955).

[26] Pitt, J. C.: 'Galileo: Causation and the Use of Geometry', in *New Perspectives on Galileo*, ed. by R. E. Butts and Joseph C. Pitt, University of Western Ontario Series in the Philosophy of Science, Vol. 14 (D. Reidel, Dordrecht, Holland, 1978).

[27] Pitt, J. C. and Tavel, M.: 'Revolutions in Science and Refinements in the Analysis of Causation', in *Zeitscrift Fur Allegaine Wissenscaftstheorie* 8 (1977).

[28] Peirce, C. S. and Tavel, M.: 'How to Make Our Ideas Clear', in *Philosophical Writings of Peirce* (Dover, New York, 1955).

[29] Popper, Karl: *Conjuctures and Refutations*, 2nd ed. (Basic Books, Inc., New York, 1965).

[30] Popper, Karl: 'Epistemology Without a Knowing Subject', in *Logic, Methodology and Philosophy of Science III; Proceedings of the Third International Congress for Logic, Methodology and Philosophy of Science, Amsterdam 1967*, ed. by B Van Rootselaar and J. F. Staal (North-Holland, Amsterdam, 1968).

[31] Popper, Karl: *The Logic of Scientific Discovery* (Hutchinson, London, 1963).

[32] Quine, W. V. O.: 'Natural Kinds', *Ontological Relativity and Other Essays* (Columbia University Press, New York, 1969).

[33] Quine, W. V. O. and Ullian, J.: *The Web of Belief* (Random House, New York, 1970).

[34] Ramsey, F. P.: *The Foundations of Mathematics and Other Logical Essays*, ed. by R. B. Braithwaite (Routledge & Kegan Paul, London, 1950).

[35] Rudner, R.: *Philosophy of Social Science* (Prentice-Hall, New York, 1966).

[36] Salmon, W. C.: *The Foundation of Scientific Inference* (University of Pittsburgh Press, Pittsburgh, 1966).

[37] Salmon, W. C.: 'Statistical Explanation', in *Pittsburgh Studies in Philosophy of Science*, Vol. 4, ed. by R. Colodny (University of Pittsburgh Press, Pittsburgh, 1970).

[38] Scheffler, I.: *Anatomy of Inquiry* (Alfred A. Knopf, New York, 1963).

[39] Sellars, W.: 'Are There Non-Deductive Logics?', in *Essays in Honor of Carl Hempel*, ed. by N. Rescher (D. Reidel, Dordrecht, Holland, 1970).

[40] Sellars, W.: 'Counterfactuals, Dispositions and Causal Modalities', in *Minnesota Studies in Philosophy of Science*, Vol. 2, ed. by H. Feigl, M. Scriven, and G. Maxwell (University Minnesota Press, Minneapolis, 1958).

[41] Sellars, W.: 'Empiricism and the Philosophy of Mind', in *Science, Perception and Reality* (Routledge & Kegan Paul, London, 1963).

[42] Sellars, W.: 'Induction as Vindication', *Philosophy of Science* **31** (1964).
[43] Sellars, W.: 'Inference and Meaning', *Mind* **62** (1953).
[44] Sellars, W.: 'The Language of Theories', in *Science, Perception and Reality* (Routledge & Kegan Paul, London, 1963).
[45] Sellars, W.: 'Philosophy and the Scientific Image of Man', in *Science, Perception and Reality* (Routledge & Kegan Paul, London, 1963).
[46] Sellars, W.: 'Science Realism or Irenic Instrumentalism', in *Philosophical Perspectives* (Thomas, Springfield, 1967).
[47] Sellars W.: 'Some Reflections on Language Games', in *Science, Perception and Reality* (Routledge & Kegan Paul, London, 1963).
[48] Sellars, W.: *Science and Metaphysics* (Routledge & Kegan Paul, London, 1968).
[49] Sellars, W.: 'Theoretical Explanation', in *Philosophical Perspectives* (Thomas, Springfield, 1967).

INDEX OF NAMES

159

INDEX OF SUBJECTS